Field Guide to the Lichens of White Rocks

Field Guide to the Lichens of White Rocks

(BOULDER, COLORADO)

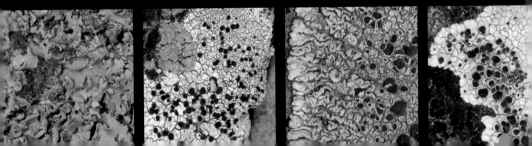

Erin A. Tripp

UNIVERSITY PRESS OF COLORADO
Boulder

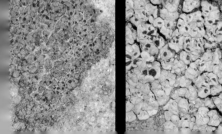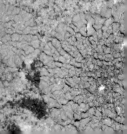

Field Guide to the Lichens of White Rocks

Introduction

The southern Rocky Mountains and adjacent prairies of Colorado represent a biological mosaic of environments typical of many montane areas of North America, especially western North America. Boulder County alone ranges in elevation from ca. 5,000 ft. to 14,000 ft., traversing one of the greatest elevational gradients of any single county in North America and hosting a range of habitats including mixed grass prairies with tallgrass relicts, submontane forested foothills, and alpine environments above treeline. Among these dominant vegetation zones are patches of rarer habitat such as geological outcroppings of sandstone or shale, eastern woodland relict forests, and fens. This book is documentation of a sandstone outcropping within the city limits of Boulder. The lichen biota of White Rocks represents an assemblage of species from the High Plains and mesas of Colorado, as well as from mid- to low-elevation montane habitats throughout the Rocky Mountains. As such, many species in this Field Guide are encountered commonly throughout central and western North America. There exist several additional Fox Hills outcrops in Colorado and neighboring states as well as sandstone outcrops of other geological time periods. The present guide will be especially useful in helping to identify the lichen constituents of those formations. Importantly, this guide treats fully the diverse crustose lichen biota in addition to the more conspicuous macrolichens.

White Rocks represents an ~100-acre ecologically important protected area within Boulder. Its biological significance is attributable in part to its geological history, climatological history, and degree of preservation but also to the fact that it is a biodiversity reservoir within a sea of agriculture and urban development (i.e., the Boulder-Denver-Longmont urban triangle). White Rocks is a rare and fragile outcropping of sandstone that rises directly above the northern margin of Boulder Creek. The outcrop itself consists of a large one- to two-tier sandstone shelf with horizontal and vertical exposed surfaces. It is approximately two-thirds of a mile in length oriented in an east-west manner. This outcropping is flanked by a more minor, adjacent sandstone exposure directly to the east, which is approximately one-half of a mile in length. White Rocks belongs to the Fox Hills Laramie Formation, dating to ca. 67 million years before present. The sandstone at

White Rocks is, as the name implies, very white in color and is composed primarily of quartz with small amounts of montmorillinite clay. The sandstone is extremely fragile and susceptible to weathering by foot travel or natural phenomena such as strong rains or high winds, but its erosion is slowed substantially by "case hardening" of the rock, which derives from hardened clay strengthened by biotic crusts—primarily lichens.

Despite the relatively small geographical size of White Rocks, the preserve is known to harbor numerous common as well as rare vascular plants and animals (Byars 1936; Weber 1949; Clark, Crawford, and Jennings 2001). This relates to the high microhabitat diversity represented at White Rocks, which is attributable to small-scale variation in relative humidity and available water, exposure to wind and sun, mineral content, aspect and steepness of slopes, and the biotic environment itself. White Rocks similarly hosts a community of common lichens (seen throughout the High Plains and Rocky Mountains) as well as rare lichens that are un- or underrepresented in Boulder County or much of Colorado. The latter builds upon prior discoveries of rare or unusual lichens present at other sandstone outcrops in North America (Skorepa 1973; Showman 1987).

Several species at White Rocks warrant conservation protection. A few may even deserve protection under the federal Endangered Species Act (ESA). However, at present, lichens are more or less excluded from federal conservation measures (only two species are currently protected by the ESA). Rare lichens at White Rocks do, however, receive local protection through the Open Space and Mountain Parks (OSMP) conservation practices. To protect the many sensitive natural resources that occur at this site, including federally regulated bird nesting habitat, White Rocks can be accessed only through permitted research and scheduled educational tours and is otherwise closed to the public(additional information on public access is available on the OSMP website www.bouldercolorado.gov/osmp).

Although a history exists of research and general interest in White Rocks Open Space, no inventory or assessment of lichens of this unique outcropping has been conducted. Thus, the primary objective of this project was to conduct a comprehensive inventory of the lichens of White Rocks. This inventory builds baseline information about the biodiversity of this important preserve as well as similar sandstone formations across western North America, enables long-term conservation planning and resource management in a data-driven manner, facilitates future lichen taxonomic and ecological research, and improves our capacity to educate the public about the importance of lichens in urban environments.

Finally, while the total lichen biota of Colorado is expected to be particularly rich given the mosaic of environments and sharp elevational and climatological gradients, a comprehensive account of Colorado lichens is lacking. Shushan and Anderson (1969) presented a lichen checklist for the state, but this list represents a small fraction of the state's total lichen biodiversity, is based entirely on literature reports, and is outdated taxonomically. The manuscript from which this Field Guide draws (Tripp 2015) is based on new field collections and adds to a list of important regional inventories in western North America that, together, will help scientists stitch together a better understanding of lichenology of the Great American West. Most immediately, this Field Guide and associated publication provide the initial steps toward a revision of the lichen biota of Colorado. Readers should refer to Tripp (2015) for more extensive information and background on White Rocks, as well as Tripp and Lendemer (2015) for descriptions of two new species from the site.

Why Lichens?

Every component of an ecosystem functions in some way vital to that ecosystem (Braun 1950). Lichens are for the most part Ascomycete fungi (the "mycobiont") with an obligate symbiotic relationship with one or more green algae or cyanobacterium (the "photobiont"). Lichens are among the most diverse and ecologically important obligate symbioses and represent important components of terrestrial ecosystems worldwide (Hawksworth 1991; Brodo et al. 2001; Cornelissen et al. 2007). In some regions of the world, lichens (together with bryophytes) contribute more to the total biotic diversity than do vascular plants (Kantvilas 1990; Jarman and Kantvilas 1995). In the relatively arid state of Colorado, the ratio of flowering plant species to lichens probably ranges between ~3 to 1 and ~3 to 2, indicating the importance of lichens to the total biota of the state. However, unlike the ±3,000 species of flowering plants in Colorado, we have only the most rudimentary knowledge of the identity, let alone ecology, of the ±1,000 to 2,000 species of Colorado lichens. Yet the pace of new species description of North American lichens far exceeds the pace of description of new plants—despite the fact that major, large-scale plant floristic projects that would seem to correlate with new species discovery exist or have been recently completed (e.g., Flora of North America Editorial Committee 1993; Jepson Flora Project 2016; Conquista et al. 2013; Weakley 2015) whereas large-scale efforts in lichenology are few (Nash et al. 2002, 2004, and 2007 represents one of the few modern lichen floras in North America).

Just like plants and animals, lichens grow and respire; just like plants, lichens also photosynthesize because of the presence of the photobiont. In addition, lichens with a cyanobacterial photobiont and/or specialized lineages of endolichenic bacteria contribute to the "fixation" of nitrogen— that is, the process by which atmospheric nitrogen is converted into a form usable by living organisms. At White Rocks, only two genera of lichens harbor cyanobacteria: *Enchylium* and *Lichinella*. Elsewhere in Boulder County, cyanolichens include *Leptogium*, *Lobaria*, *Nephroma*, *Peltigera*, and, if you are really lucky, *Sticta*. Genomic investigation of bacterial communities of lichens is in its infancy, but it is likely that other nitrogen-fixing prokaryotes in addition to cyanobacteria are far more prevalent in lichens than previously recognized (Kane, Tripp, Lendemer, and McCain, in progress).

Lichens that grow on rock also influence nutrient distribution in their environments in another manner: as primary decomposers of parent rock material into what will ultimately become soil. The lichen mycobiont manufactures and secretes various chemicals that aid fungal hyphae in the penetration of rock surfaces. Once below the surface, hyphae grow among the crystalline structures and between rock cleavage lines, further fracturing the parent material. As lichens grow and decompose, they also trap fine soil particles among their surfaces, encouraging saprophytic fungi and bacterial growth, further adding to the process of soil formation. Lichens that grow on trees influence local distribution of nutrients by trapping airborne nutrients including water, thereby helping to maintain relative humidity as well as providing shelter for microorganisms. Whether their presence is detrimental to living substrates (e.g., trees) is a question unanswered. On the whole, however, it is widely appreciated that healthy ecosystems have ample lichen cover, and unhealthy ecosystems lack lichen cover. A stop along a forest bordering northern stretches of Interstate 95 will demonstrate the latter. A hike at 13,000 feet in Rocky Mountain National Park will demonstrate the former.

Beyond their roles in ecosystem nutrient cycling, lichens also function prominently in many food webs. In Colorado, lichens figure substantially in the diets of both large- and small-bodied animals ranging from elk to moose, deer, squirrels, birds, snails, mites, and insects (Sharnoff 1994; Pettersson et al. 1995). In boreal ecozones, lichens serve as primary food sources for mountain goats, and caribou cannot live without them. Lichens are also used by numerous animals in nest construction (Ladd 1998) or medicinally in some human societies (Wei et al. 1982). In fact, lichens manufacture hundreds of secondary compounds not known elsewhere in nature and whose ecological functions are at best minimally understood. This chemical diversity has (perhaps unsurprisingly) been exploited by lichenologists as a source of taxonomic information: thin layer chromatography (TLC) to determine lichen secondary chemistry is necessary to identify at least a third of all North American species with confidence.

Finally, lichen abundance and species richness have long been appreciated as indicators of the richness of other taxa, habitat quality, and air quality of a particular region (De Wit 1983; Nilsson et al. 1995; Bergamini et al. 2007; Tripp and Lendemer 2012; Lendemer et al. 2013). Nordén and colleagues (2007) found a significant correlation between temperate and deciduous forest lichen, bryophyte, and wood fungi diversity and the number of rare, Red Listed species in these groups. In the southeastern United States, McCune and colleagues (1997) found lower lichen diversity and abundance in areas of higher air pollution. Great Smoky Mountains National Park, which contains some of the most extensive tracts of old growth forest remaining in eastern North America, is lichenologically the most diverse park in the United States and contains upward of half of all species present in eastern North America (Tripp and Lendemer 2012; Lendemer et al. 2013). For the most part, relatively common species have been used as forest indicators (McCune et al. 1998). However, recent studies are also bringing new recognition to the rarer component of the lichen biota, and inventories have indicated that occurrence of rare plants and animals correlates to occurrence of rare lichens (Lendemer et al. 2013).

Lichen Biology: The Basics

The traditional concept of the lichen symbiosis has been that the mycobiont and photobiont form a mutualist relationship, with the fungus providing a protected environment in which the photobiont can thrive and the photobiont supplying nutritional products of photosynthesis to feed the fungus. More recent perspectives suggest this traditional view may not be accurate, with one or both of the partners functioning parasitically at times (reviewed in Richardson 1999). Recent research has also demonstrated that the lichen symbiosis itself is far more complex than previously understood, with the discovery of multiple photobiont genotypes in a single organism (Muggia et al. 2013), as well as additional partners including endolichenic fungi or bacteria, whose functions are for the most part still under investigation (Arnold et al. 2009; Spribille et al. 2016).

The vegetative or non-sexual portion of the lichen body is termed a thallus (plural: thalli) and consists generally of four layers. The upper and lower layers comprise the cortex and are composed of densely packed fungal hyphae. The upper cortex is where many accessory pigments that give lichens their color reside. Just underneath the upper cortex is the photobiont layer, which varies from light to dark green to orange to blue, depending on photobiont type (e.g., coccoid green alga, *Trentepohlia*, a cyanobacterium). The layer below the photobiont is termed the medulla; it is usually bright white in cross section (but can be brightly pigmented in some species, such as in *Vulpicia pinastri*, which is common in Boulder

County and has a bright yellow medulla) and consists of loosely arranged fungal hyphae. Below the medulla, the lower cortex (where present, see below) is generally white, gray, brown, or black in color.

Lichen upper cortices are modified or ornamented in myriad ways. Specialized features include cyphellae and pseudocyphellae, maculae, perforations, cilia, and pruina. Pruina is a whitish coating on the upper surfaces of some lichens, giving an appearance of powdered sugar. Pruina is very common among species at White Rocks and can often be seen on surfaces of apothecia (e.g., *Diplotomma venusta*) and/or on surfaces of thalli (e.g., *Psora tuckermanii*). Degree of pruinosity can vary to such an extent even over a single thallus as to completely obscure the true color of the upper cortex by giving the lichen the appearance of a white color (as in *Acarospora strigata*: dark brown when epruinose but bright white when pruinose). Lichen lower cortices are differentiated by color, texture, and presence of attachment structures such as rhizines, which anchor lichens to their substrate and are typical of many foliose lichens.

Not all lichens have differentiated upper and lower cortices, and not all lichens have both cortices. First, by definition, most fruticose lichens (see growth form information below) have only one type of cortex, that is, an upper cortex is non-differentiable from a lower cortex. We have only one fruticose lichen at White Rocks: *Lichinella stipatula*, which is microfruticose. Second, by definition, crustose lichens lack a lower cortex. The majority of the lichen biota at White Rocks is crustose (e.g., *Caloplaca trachyphylla*, *Lepraria finkii*, *Acarospora* spp.). Finally, most foliose lichens have well-differentiated upper and lower cortices, but a small number of foliose lichens are characterized by having "ecorticate" lower surfaces (like crustose species), such as species in the genus *Peltigera* and *Heterodermia* (not present at White Rocks but commonly encountered elsewhere in Boulder County).

Lichen Reproduction

Lichens reproduce sexually and/or asexually. Most commonly, a given species reproduces primarily through one but not both means, that is, a species is either sexual or asexual (Tripp 2016); however, asexual species are on occasion encountered with sexual reproductive structures (see below). At White Rocks, *Psora tuckermanii* is almost always found with sexual reproductive structures, whereas *Verrucaria furfuracea* is always found with asexual reproductive structures. Just like all other Ascomycete fungi, the basic unit of sexual reproduction in lichens is the ascoma. Lichen ascoma (plural: ascomata) occur primarily in two forms: disc-shaped apothecia and flask-shaped perithecia. Inside the ascomata are sacs that contain the products of meiosis: ascospores. Spores are released from either the open discs of an apothecium (as in *Candelariella rosulans*) or from a pore-like opening at the top of a perithecium (as in *Staurothele areolata*). Because sexual reproduction involves only the fungus and not the alga, dispersed fungal ascospores must encounter a non-lichens-compatible alga (or steal an alga from a different lichen, which surely happens with some frequency) to give rise to a new thallus. Lichen apothecia, the most common type of reproductive structure, are extremely diverse morphologically. They can occur with or without thalline margins, and when present, these are sometimes ornamented. Thalline margins are typically described as lecideine if they lack algae and are carbonized or black in color (as in *Lecidea hoganii*) or lecanorine if algae are present and margins are the same color as the thallus (as in *Lecanora muralis*). Apothecia may be concave at maturity (*Xanthoparmelia coloradoensis*), convex (*Lecidella carpathica*), flat (*Candelariella clarkiae*), lirellate or lip-like (this modification not present at White Rocks), or raised on stalks or podetia (this modification also not present at White Rocks but characteristic of the diverse and widespread genus *Cladonia*). Sexual reproductive structures can range from comprising nearly the entire visible lichen (*Lecanora flowersiana*) to being scattered across the thallus (*Caloplaca sideritis*) to being rare (*Caloplaca decipiens*) or completely unknown (*Lepraria finkii*) for a given species. Finally, sexual ascospores are extremely important in lichen identification. They range from hyaline to brown at maturity and simple to transversely sepatate to muriform. A given sexually reproducing species generally has a diagnostic number of ascospores per ascus. The most typical number is eight spores per ascus, reflecting meiosis followed by a single mitosis event, but severe reductions in the number of spores per ascus are possible (as in one spore/ascus in *Rhizocarpon disporum*), as are additional mitotic events to yield many more spores per ascus (as in *Acarospora*, with thirty-two to hundreds of tiny spores/ascus). Sexually reproducing crustose lichens almost always need to be sectioned by hand and studied under a compound microscope to identify species with confidence.

Asexual reproduction in lichens occurs primarily through modifications of the thallus into specialized lichen propagules. Unlike sexual reproduction, asexual propagules generally disperse the fungus and the alga together as a unit, the exceptions being structures termed pycnidia that house asexual, fungal-only conidia (pycnidia can be seen as tiny black dots on the surface of one of the lobes in the uppermost portion of the photo of *Xanthoparmelia coloradoensis*). The most common asexual reproductive structures are termed soredia (rounded masses of hyphae surrounding one or a few algal cells, as in *Caloplaca decipiens*), isidia (columnar versions of soredia, as in *Xanthoparmelia lavicola*), and phyllidia (micro-lobules that break off of the main thallus, not present at White Rocks). Soredia usually develop in regions of the thallus known as soralia. Fragments of the lichen thallus may also break away and disperse to yield a new lichen thallus.

Growth Forms

Most lichens can be broadly characterized as having one of four major growth forms: foliose, fruticose, crustose, or squamulose (note, however, that similarity in growth form does not convey shared evolutionary relationship). A much smaller number of lichens defy placement into one of these four primary categories and are best described using more specific terms such as "leprose" (as in *Lepraria*) or "filamentous" (this form not present at White Rocks, but species in the genus *Coenogonium* are good examples). Of the four major growth forms, foliose lichens most readily exemplify the four-layered, stratified thallus described above. Foliose species are the large and readily visible lichens that come to mind when one conjures an image of macrolichens covering their substrates via flattened, "leaf-like" thalli. Various aspects of both macro- and micromorphology as well as chemistry aid in the identification of foliose lichens. For example, in Colorado, the widespread species *Parmelia saxatilis* and *Parmelia sulcata* are characterized by the network of pseudocyphellae on their upper cortices. The diverse and primarily eastern North American genus *Parmotrema* is characterized largely by the presence of cilia scattered about the margins of the thallus lobes. The most common foliose lichens at White Rocks and indeed in much of western North America are species of the genus *Xanthoparmelia*. Other genera including *Montanelia*, *Umbilicaria*, *Punctelia*, *Physcia*, and *Physciella* are common foliose lichens in the Front Range of Colorado.

Fruticose lichens are generally the most three-dimensional of the four major growth forms. They almost always have indistinguishable upper and lower cortices and are often (but not always) cylindrical or sub-cylindrical in cross section. Fruticose lichens adhere to substrates through a single or a few holdfast structures and are often seen dangling from trees and rocks. The most common fruticose lichens in Boulder County are species of *Usnea* and, at higher elevations, species of *Cetraria* and *Bryoria*. At lower elevations of White Rocks, we have only one fruticose species, *Lichinella stipatula*, considered to be "dwarf fruticose" because of its diminutive size. One of the most species-rich lichen genera in the world (and a genus of fruticose lichens), *Cladonia*, occurs at both high and low elevations in Colorado but is conspicuously absent from White Rocks. Some genera such as *Aspicilia* contain both fruticose species and species with other growth forms (mostly crustose in *Aspicilia*). Fruticose lichens together with foliose species comprise a taxonomically artificial group typically referred to as "macrolichens."

Crustose lichens are by far the most understudied of the four growth forms, even though they constitute over half of total lichen diversity in most areas worldwide. At White Rocks, over 70 percent of the total lichen diversity (41 of 58 species) is crustose. This percentage probably applies to much of Colorado as well as throughout western North

America, where crustose species are the rule rather than the exception. Crustose lichens are characterized by their lack of a lower cortex; instead of having rhizines or other attachment structures, their lower cortices are in direct contact with the substrate. Thus, crustose lichens cannot easily be removed from substrates and, as such, pieces of bark, rock, or soil serving as the lichen substrate must be removed together with the lichen for further laboratory study and museum accessioning. Many crustose species are large and seen easily with the naked eye (*Acarospora stapfiana*, *Caloplaca saxicola*); an equal number are minute and require magnification to discern even the most general aspects of morphology (*Polysporina simplex*, *Rinodina venostana*). This added challenge has contributed to less overall knowledge of crustose lichens compared with that of North American macrolichens. *Lecanora*, *Lecidea*, *Lecidella*, *Verrucaria*, and *Acarospora* are among the most diverse and important genera of crustose lichens in western North America, including at White Rocks; they can be found abundantly on rock and, to a much lesser degree, on other substrates.

The squamulose growth form is intermediate between foliose and crustose growth forms. Squamulose species are characterized as having miniature and often overlapping lobes or lobules. Because of the small size of these lobes, they superficially resemble (and are often discussed together with) crustose lichens. The squamulose growth form is very common in western North America and appears to be correlated with growth on soil or loose rock substrates. Examples of squamulose genera at White Rocks include *Endocarpon*, *Psora*, *Placidium*, and one species of *Lichinella* (*L. nigritella*).

Substrates

Lichens occur on a tremendous variety of substrates ranging from the bark of living plants (corticolous) to rock (saxicolous), soil (terricolous), decomposing wood (lignicolous), and leaf surfaces (foliicolous). Rarer substrates include tree resin (resinicolous), rotting metal (metalicolous), and the exterior surfaces of animals such as tortoises (zooicolous). Saxicolous lichens can generally be sorted into calcareous-loving and non-calcareous-loving species, that is, species that grow on limestone or other calcium-rich rock versus species that grow on siliceous rock such as granite or sandstone. A given species of corticolous lichen will typically grow on either acid bark (mostly conifers) or more basic to neutral bark (mostly hardwoods). Hardwoods almost always host more diverse lichen communities than conifers. Lichens that occupy decaying or non-living wood such as old stumps, fence posts, or signs are mostly specific to these types of substrates; that is, they typically do not also occur on living wood. In arid portions of western North America, saxicolous followed by terricolous and lignicolous substrates are the most important growing surfaces for lichens.

Additional Remarks

Ecological functions of lichens include contributions to biogeochemical cycling, biomass production, pollutant sequestration, decomposition, soil formation, and habitat or nutrition sources for an untold diversity of organisms (Szczepaniak and Biziuk 2003; Cornelissen et al. 2007; Bobbink et al. 2010). This functional diversity coupled with impressive ecological amplitude contributes prominently to the conspicuous diversity and abundance of lichens across the globe, even in extreme environments where few other organisms occur (Lutzoni and Miadlikowska 2009). Yet the diversity and ecological relevance of lichens have gone unnoticed by a great number of natural history enthusiasts, in part because of a lack of inclusion of lichen biology in science curricula in North America. It is hoped that the present contribution will help remedy this problem by providing an accessible, user-friendly guide to educators, conservationists, and land managers as well as lichenologists. Above, the reader is equipped with basic information on lichen morphology, reproduction, and ecology to enable a general understanding and appreciation of the species represented in this Field Guide.

About the Guide

The above text is intended to provide only the most basic introduction to lichen biology. For a much more comprehensive, general overview of lichen biology, readers are encouraged to read the excellent introductory chapters of Brodo and colleagues (2001) as well as introductory pages in Nash and colleagues (2002, 2008).

The following species pages are intended to capture fully all currently known lichens present at White Rocks. Thus, all fifty-six known species are included, as is a dichotomous key to assist with their identification. For each species in the Field Guide, a general description is provided, which is intended to serve the primary purpose of conveying easy-to-recognize features of a given lichen as well as distinguishing it among close relatives, either at White Rocks or in the surrounding area. Additional information regarding ecology, substrate, and distribution and a primary literature reference to further information about the species are provided. Spot tests refer to standard chemical reagent assays (Brodo et al. 2001). Some use of technical terminology was necessary in preparing the text, and as such, a truncated glossary to a small selection of lichenological terms used in this book can be found in the back of this guide.

All collections were photographed in the field by the author (except where noted) using a Nikon D7100 digital SLR with a 105 mm 1:1 macro-lens and ring flash. The primary set of voucher specimens has been deposited at the University of Colorado Museum of Natural History (COLO Herbarium), and duplicates are deposited elsewhere, primarily the New York Botanical Garden (NY Herbarium). Photographs of all species included in this Field Guide derive from the White Rocks field setting without exception, despite the fact that in several instances I have on hand better-quality images for a given species. The rationale for doing this is that the primary purpose of this Field Guide is to facilitate future research, conservation, and management of the White Rocks lichen biota, and providing photographs of species in essentially a raw, unedited state conveys as much of the general lighting and site-specific features of a species as possible. Thus, the degree of thallus pruinosity of *Acarospora strigata* at

White Rocks is depicted, rather than pruinosity typical of the species as it occurs elsewhere.

This Field Guide is somewhat unique among professional or amateur field guides in that photographs of all species are in all cases tied to a physical museum (herbarium) voucher specimen, thus enabling future researchers to confirm or re-investigate the identification of a given voucher specimen at any point in time. This guide utilizes modern taxonomies of species arrangements, largely following that of Esslinger's (2015) North American Lichen Checklist, with rare exception. A translucent box superimposed on each photograph depicts the museum voucher(s) upon which the photograph was made. The number refers to my herbarium collection numbers. Credit to the photographer(s) is provided in each translucent box (following the voucher), which is important in any case but especially in lichenology given that in some cases it takes thirty bad photographs before one decent photograph is made.

The present study serves as the first documentation of the lichen biota of a rare sandstone formation in Boulder, Colorado. Eighty-two collections yielded 57 total species (see the appendix for a list of species with taxonomic authorities). Somewhat remarkably, there have been very few lichen inventories of sandstone formations across North America. By far the most relevant to the present study was that of Anderson (1962), who inventoried the Dakota sandstone formation of northern Colorado. In that study, Anderson documented 130 species over a stretch of ca. sixty-five miles, ranging from Boulder northward along the Front Range of the Rocky Mountains (width of band not reported in the publication but likely not more than fifteen to thirty miles wide). The White Rocks formation supports just under half of the total number of species in Anderson's study but occupies only 100 acres versus ca. 832,000 acres (65 × 20 miles, or 1,300 mi^2), highlighting the ecological significance of White Rocks Open Space. Moreover, the White Rocks lichen biota contains at least two (i.e., *Candelariella clarkiae*, *Lecidea hoganii*) and as many as four species new to science (Tripp and Lendemer 2015; Tripp 2015), as well as one newly confirmed report for the United States (*Rinodina venostana*; Tripp 2015). My hope is that this Field Guide improves management practices at

White Rocks and other urban open spaces and helps educate staff as well as the public about the importance of lichens in our environments.

I an indebted to Darrin Pratt, Daniel Pratt, Laura Furney, Cheryl Carnahan, and Jessica d'Arbonne at the University Press of Colorado for extensive assistance in preparing this Field Guide for publication. My most sincere thanks to Lynn Riedel of the City of Boulder's Open Space and Mountain Parks Program for her enthusiasm toward the lichen biota of White Rocks and for facilitating all aspects of the research that enabled this Field Guide, including permitting and partial funding. Thanks to Dina Clark, Vanessa Díaz, and James Lendemer for contributions to fieldwork and photography. I thank Dina Clark, Sue Hirschfeld, and Bob Weimer for permission to use photographs in the introductory pages. I am indebted to the following individuals who helped confirm identifications for select collections: Othmar Breuss, Ted Esslinger, Kerry Knudsen, James Lendemer, and John Sheard. Sue Hirschfeld offered welcome information on the geological history of White Rocks. Collections Managers Tim Hogan and Dina Clark, Project Manager Ryan Allen, and student interns at the University of Colorado Herbarium (Vanessa Díaz, Joeseph Kleinkopf, Breanna Leinbach, and Melissa Smithson) assisted with the processing and curation of specimens from the inventory and contributed some of the common names in this guide. I am indebted to two anonymous reviewers for contributing helpful peer reviews to an earlier version of this guide. Funders who provided financial support include the University of Colorado Museum of Natural History, the Colorado Native Plant Society Board of Directors and contributing members, and the City of Boulder, Open Space and Mountain Parks. Finally, I thank Dr. Bill Weber for saving White Rocks for me, to study and to learn from.

FIGURE 1.1. Incoming storm over the city of Boulder, as seen from White Rocks (August 2014). Photo by Dina Clark.

Field Guide

FIGURE 1.2. Boulder atop upper shelf of sandstone formation showing four common lichens at White Rocks: *Caloplaca trachyphylla* (orange) parasitized by *Acarospora socialis* (pale chartreuse toward center of *Caloplaca* thalli, seen especially toward the right-hand side), *Acarospora stapfiana* (white), and *Candelariella rosulans* (yolk yellow). Photo by Sue Hirschfeld.

FIGURE 1.3. Aerial photo of White Rocks in 1975 with eastern plains in the background, prominent sandstone outcropping in the western foreground, and Boulder Creek flanking southern edge of formation. Photo by Bob Weimer.

Acarospora obpallens

Acorn Cups

Acarospora obpallens is an attractive crustose species with irregularly rounded areoles often somewhat dispersed across the substrate. Thalli are generally the color of "coffee with milk," that is, pale brown to yellowish-brown. Apothecial discs are reddish-brown in color and range from epruinose to heavily pruinose (as seen at left). This species is morphologically easily distinguishable from other *Acarospora*. When in doubt, do a C test.

CHEMISTRY: Gyrophoric acid (major), lecanoric acid (minor), 3-hydroxygyrophoric acid (trace), and methyl lecanorate (trace); spot tests: K–, C+ red (cortex), KC+ red (variable), P–, UV–.

SUBSTRATE AND ECOLOGY: Terricolous or saxicolous on sandstone, volcanic rock, or decomposing granite; occasionally on other lichens.

DISTRIBUTION: *Acarospora obpallens* is an excellent species for your "lichen list" because it is common throughout central and western North America.

LITERATURE: Knudsen, K. 2007. "*Acarospora*." In T. H. Nash III, B. D. Ryan, C. Gries, and F. Bungartz, eds., *Lichen Flora of the Greater Sonoran Desert Region*, vol. 3. Tempe: Lichens Unlimited, Arizona State University, 1–38.

#4830 (PHOTO: E. TRIPP)

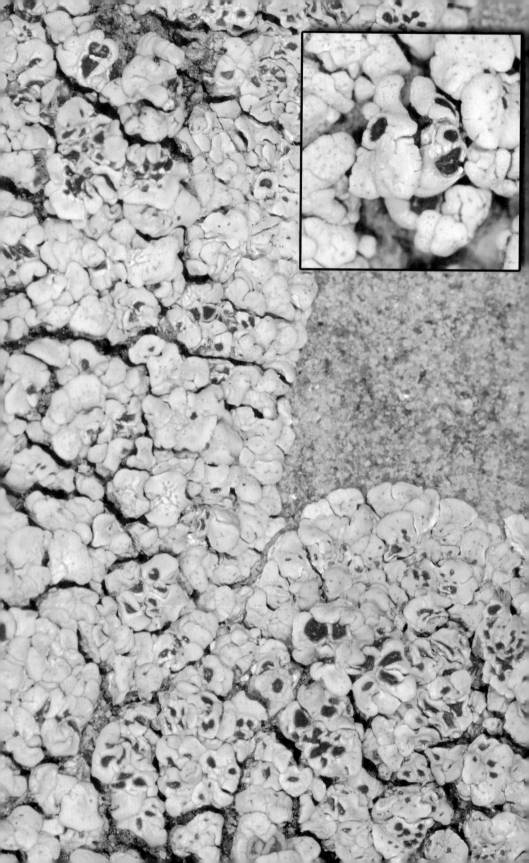

Acarospora stapfiana

Corn Pops

Acarospora stapfiana is a spectacular areolate to squamulose crustose lichen that is extremely variable and wide ranging in western North America. Of the yellow Acarosporas in western North America, *A. stapfiana* is by far the most common. In our area, it can be distinguished from *A. schleicheri* by a lack of imbricate areoles (common in *A. schleicheri*) and its saxicolous to lichenicolous habit (versus primarily terricolous in *A. schleicheri*). Rather than other Acarosporas, *Acarospora stapfiana* is most likely to be confused with another bright yellow saxicolous species that is common throughout Boulder County, *Pleopsidium flavum,* but the latter is usually effigurate; has a brighter, shiny yellow thallus; has smaller, yellower apothecia (< 1 mm in diam.), and has a K/I+ tholus.

CHEMISTRY: Rhizocarpic acid, +/− epanorin in trace amounts; spot tests: K–, C–, KC–, P–, UV+ orange.

SUBSTRATE AND ECOLOGY: Very common on sandstone, volcanic, or granitic rocks, rarely on soil or other substrates. *Acarospora stapfiana* is often a pioneer of new habitats and is frequently seen growing in the center of or near thalli of *Xanthomendoza trachyphylla* (as in the photo; see far left-hand side), suggesting that it may be opportunistically parasitic.

DISTRIBUTION: Widespread in western North America and Mexico.

LITERATURE: Knudsen, K. 2007. *"Acarospora."* In T. H. Nash III, B. D. Ryan, C. Gries, and F. Bungartz, eds., *Lichen Flora of the Greater Sonoran Desert Region*, vol. 3. Tempe: Lichens Unlimited, Arizona State University, 1–38.

#4806 (PHOTO: E. TRIPP)

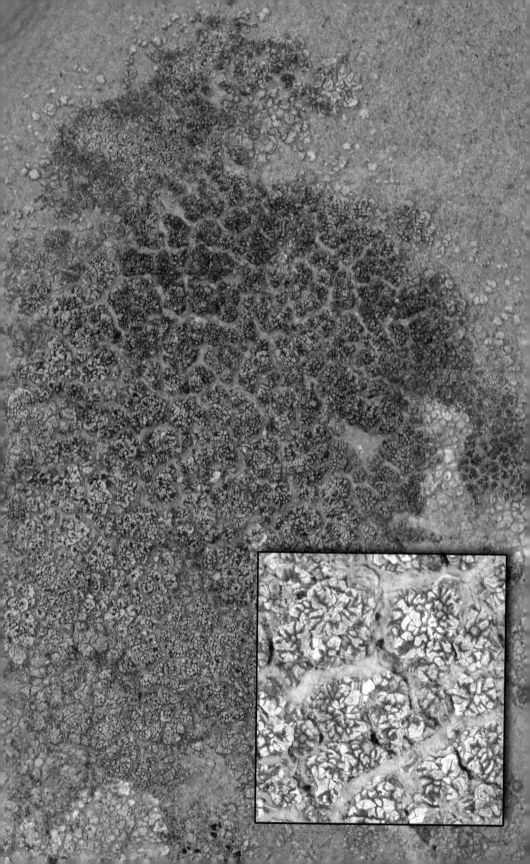

Acarospora strigata

Frosty Fissures

Acarospora strigata is an areolate to verruculose crustose lichen that is extremely variable in appearance because of the white pruina that can range from dense to entirely lacking across a single thallus (see photo). When epruinose, thalli of *A. strigata* are generally brown in color; when pruinose, thalli range from white to bluish-gray. Apothecial discs of this species are reddish-brown in color. *Acarospora strigata* can be recognized by its radial fissures that traverse the thallus and surround apothecia, giving the lichen a cracked appearance. Its broadly ellipsoid spores help distinguish it from *A. veronensis.*

CHEMISTRY: No substances; spot tests: K–, C–, KC–, P–, UV– to occasionally appearing pale green or orange because of mineral inclusions.

SUBSTRATE AND ECOLOGY: Common on granite or limestone, and especially successful in harsh environments.

DISTRIBUTION: Widespread in central and western North America (also reported from China, Siberia, Mexico, and South America).

LITERATURE: Knudsen, K. 2007. "*Acarospora.*" In T. H. Nash III, B. D. Ryan, C. Gries, and F. Bungartz, eds., *Lichen Flora of the Greater Sonoran Desert Region*, vol. 3. Tempe: Lichens Unlimited, Arizona State University, 1–38.

#4868 (PHOTO: E. TRIPP)

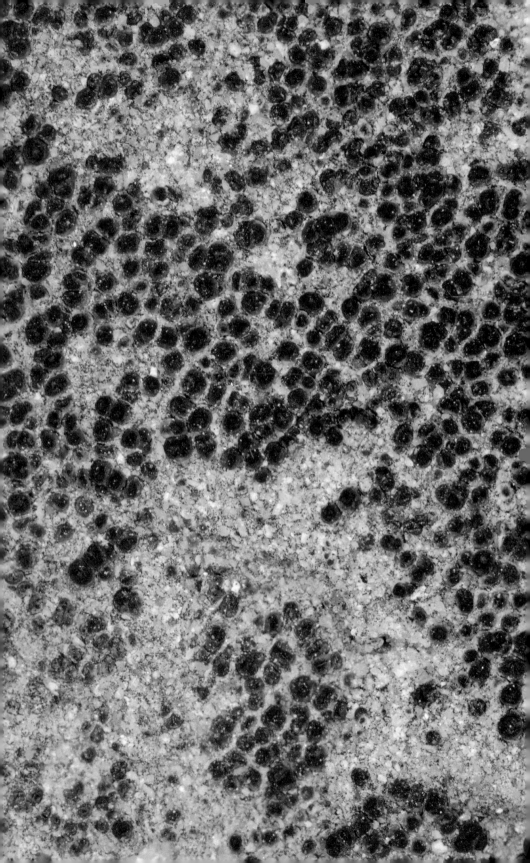

Acarospora sp. nov.

Gothic Pom-Poms

This is a curious species of *Acarospora* (readily confirmed to genus by its polysporous asci) whose identity is still being investigated; we think it is new to science and will soon describe it as such. Alas, with or without a name, it belongs in this book because it is readily distinguished from all other species in the genus at White Rocks by its tiny, rounded areoles with dark brown to almost black apothecia. Of the two Acarosporas with brown thalli at White Rocks, *A. strigata* has thalli with radial fissures and discs that are reddish-brown in color, and *A. obpallens* has much larger, paler yellowish-brown areoles.

CHEMISTRY AND SPOT TESTS: Unknown.

SUBSTRATE AND ECOLOGY: Very rare on open, exposed sandstone.

DISTRIBUTION: Unknown outside of White Rocks.

LITERATURE: None.

#4820 (PHOTO: E. TRIPP)

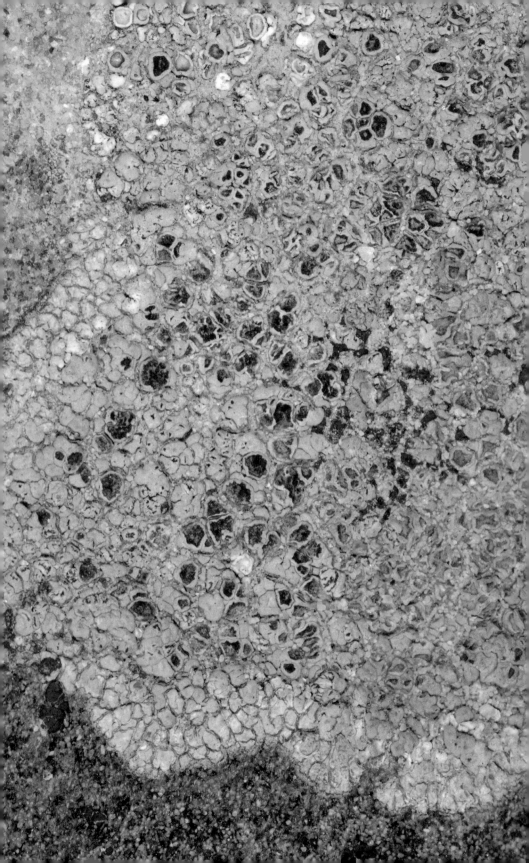

Aspicilia cinerea

Stool Pigeon

Aspicilia cinerea is a crustose species with areoles that are generally angular. The species is characterized by its gray to white thallus and its large apothecia with black discs that are distinctly elevated from the thallus (other *Aspicilia* in our area have discs that are sunken within the thallus). *Aspicilia cinerea* can be found with or without a black to bluish-black prothallus (present at left), and, like all members of the genus, it has apothecia with lecanorine margins. In our area, *A. cinerea* is most likely to be confused with *A. caesiocinerea,* which differs most conspicuously by its chemistry (the latter is K–, P–, with aspicilin [and lacking norstictic acid]). Beginners sometimes confuse *Aspicilia* with the genus *Acarospora,* but the two are easily distinguishable by the more continuous thallus and non-polysporous asci in *Aspicilia* (versus more areolate thalli and polysporous asci in *Acarospora*).

CHEMISTRY: Norstictic acid (major), connorstictic acid (trace); spot tests: K+ red (cortex, medulla), C–, KC–, P+ orange (cortex, medulla), UV–.

SUBSTRATE AND ECOLOGY: Common on silica-rich rocks, schist, and volcanic rock; rare on calcareous rock.

DISTRIBUTION: Widespread in North America except absent from the southeastern United States (also in northern regions of Europe and Asia).

LITERATURE: Brodo, I. M., S. D. Sharnoff, and S. Sharnoff. 2001. *Lichens of North America.* New Haven, CT: Yale University Press.

#4880 (PHOTO: V. DÍAZ)

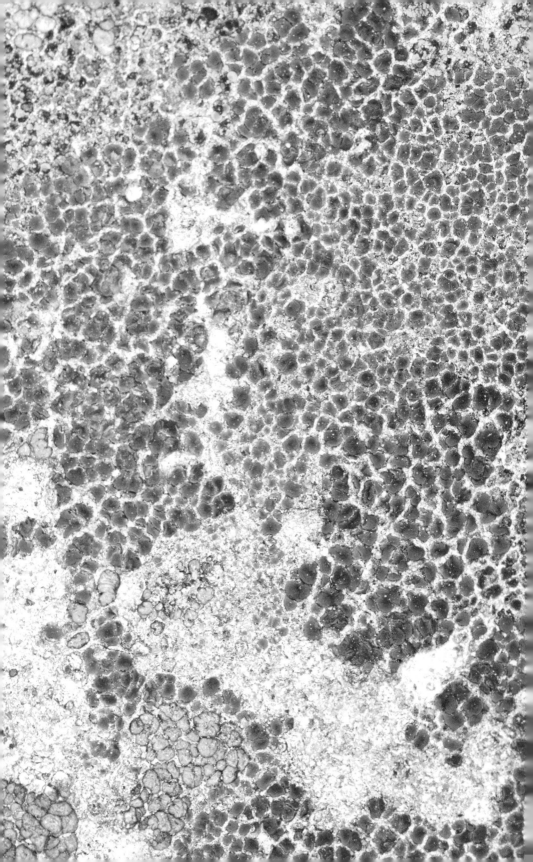

Caloplaca atroflava

Rocky Mountain Sunburn

Caloplaca atroflava is easily differentiated from all other species of *Caloplaca* at White Rocks by its prominent, orangish-red apothecia and lack of an obvious thallus. Okay, let's be honest: it actually has a conspicuous, albeit very thin, gray thallus, mostly associated with older fruiting bodies. You will no doubt first see the apothecia, but bring a 20× hand lens, open your mind, and peer into this wonderful microcosm. Then take a step back, and notice how the background color changes from nearer to further from fruiting bodies. Elsewhere in our area, this species is most likely to be confused with *C. epithallina*, which similarly has dark orangish-red apothecia, but *C. epithallina* truly lacks a thallus and is always lichenicolous (parasitic) on other lichens. *Caloplaca grimmiae* is also parasitic and lacks a thallus but has apothecia with darkened margins (versus margins concolorous with discs in *C. epithallina*). If you are lichenizing higher in elevation in Boulder County, don't confuse *C. atroflava* for *C. arenaria*.

CHEMISTRY: Sedifolia green; spot tests: K+ red (apothecia margin, epihymenium), K+ fleeting dull purple (thallus, visible only in water mount), C+ red-brown (apothecia margin), KC+ red-brown (apothecia margin), P–, UV–.

SUBSTRATE AND ECOLOGY: Restricted to exposed, silicaceous rocks—like those at White Rocks.

DISTRIBUTION: Known primarily from Arizona to California. This report might represent the easternmost population of this species.

LITERATURE: Wetmore, C. M. 2007. "Caloplaca." In T. H. Nash III, B. D. Ryan, C. Gries, and F. Bungartz, eds., *Lichen Flora of the Greater Sonoran Desert Region*, vol. 3. Tempe: Lichens Unlimited, Arizona State University, 179–220.

#4824 (PHOTO: E. TRIPP)

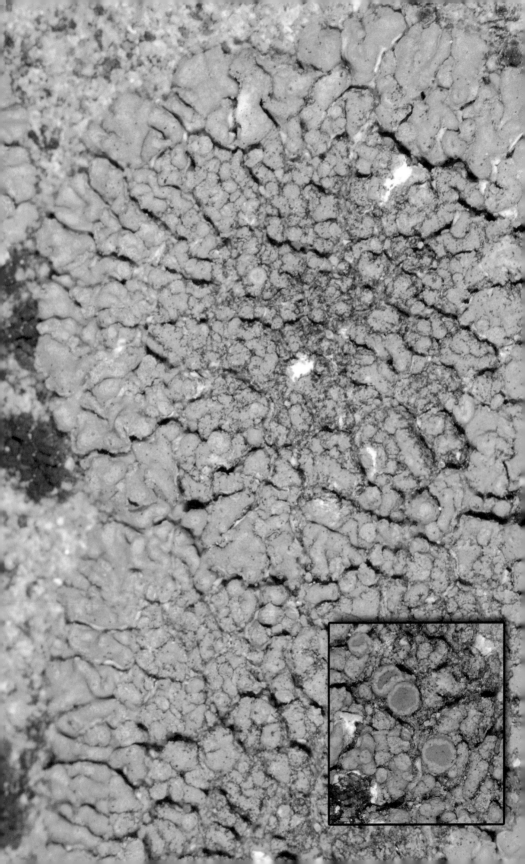

Caloplaca decipiens

Broken Yokes

Caloplaca decipiens is a common and easily recognized crustose species with radiating, convex lobes that are crowded toward the tips. One of the most notable features of this species is its more orangish-yellow coloration, which contrasts markedly with most other species in the genus, which are typically deep orange to orange-red in color. Unlike most other species of *Caloplaca* in our area, which reproduce sexually, *C. decipiens* reproduces primarily asexually through soredia at the lobe tips (see main photo; as a side note, over 75 percent of the ~170 species of *Caloplaca* present in North America reproduce sexually [Tripp and Lendemer, unpub. data]). *Caloplaca decipiens* is much more rarely found with sexual reproductive structures that, when found, co-occur with soredia. This dual mode of reproduction applies to populations at White Rocks in 2014 (see fruiting bodies in the inset photo) and certainly seems like a winning strategy to me. The combination of soredia plus thallus color makes *C. decipiens* unmistakable in our area.

CHEMISTRY: Emodin, fallacinal, parietin, teloschistin; spot tests: K+ red (apothecia margin), K+ violet (cortex), C–, KC–, P–, UV–.

SUBSTRATE AND ECOLOGY: Extremely common on non-calcareous and calcareous rocks (throughout the city of Boulder, this species can be found as a primary pioneer of artificial rock walls).

DISTRIBUTION: Widespread in central and western North America (also in Europe).

LITERATURE: Wetmore, C. M. 2007. "Caloplaca." In T. H. Nash III, B. D. Ryan, C. Gries, and F. Bungartz, eds., *Lichen Flora of the Greater Sonoran Desert Region*, vol. 3. Tempe: Lichens Unlimited, Arizona State University, 179–220.

#4845 (PHOTO: J. LENDEMER)

Caloplaca pratensis

K+ Crust

Caloplaca pratensis is an overlooked crustose lichen with a thin, continuous to areolate gray thallus (figure at left shows a thallus that approximates a grain of rice in total length). But let's not discriminate. Superficially, the species is an atypical member of the genus *Caloplaca* because of its gray thallus (most are brightly colored; see others in this Field Guide). Closer inspection, however, will reveal K+ purple reactions in the soredia and epihymenium, as well as polarilocular spores, both features typical of the genus. Like *C. decipiens*, *C. pratensis* is somewhat unusual because it can sometimes be found with both sexual (i.e., apothecia) and asexual (e.g., soredia) reproductive structures. The species is occasional at White Rocks where it is unlikely to be confused with any other species because of the above characteristics. The only other gray member of the genus present at White Rocks, *Caloplaca sideritis*, is very distinctive morphologically because of its orange apothecial discs with a dark excipular ring; furthermore, the discs of *C. sideritis* are raised above the thallus (versus brownish-black discs that are more embedded in the thallus and lack a dark excipular ring in *C. pratensis*). Elsewhere in Colorado and on similar substrates, *C. pratensis* might be confused with *C. alboatra*, which is non-sorediate. See Wetmore (2009) for additional comparisons to other species of *Caloplaca*.

CHEMISTRY: Thalloidima green; spot tests: K+ violet (soredia, epihymenium, thallus), C−, KC−, P−, UV−.

SUBSTRATE AND ECOLOGY: Most commonly on concrete or other calcareous surfaces and rarely on sandstone, as in the White Rocks collection.

DISTRIBUTION: Relatively common in the north-central Great Plains, extending into eastern Colorado.

LITERATURE: Wetmore, C. M. 2009. "New Species of *Caloplaca* (Teloschistaceae) from North America." *Bryologist* 112: 379–86.

#4836 (PHOTO: J. LENDEMER)

Caloplaca saxicola

Flaming Cowgirls

Caloplaca saxicola is a common crustose species recognized at White Rocks by its definitively but not too extravagantly lobed margins. Its orange color plus this lobing make *Caloplaca saxicola* easily distinguishable from all other members of the genus at White Rocks except for *C. trachyphylla*, which has much longer lobes (2–5 mm versus 1–2 mm in *C. saxicola*) and bigger thalli that are also lumpy in appearance. The other species at White Rocks with which *C. saxicola* is potentially confusable is *Xanthoria elegans*. The latter is one of the most abundant orange lichens on rock in Colorado as a whole, but it can be separated from *C. saxicola* by its foliose thallus with a clearly differentiated lower cortex, unlike members of *Caloplaca* that are crustose and lack a lower cortex and thus cannot be removed intact from the substrate.

CHEMISTRY: Emodin, fallacinal, parietin, teloschistin; spot tests: K+ red (thallus, apothecia margin), C–, KC–, P–, UV–.

SUBSTRATE AND ECOLOGY: Very common on non-calcareous and calcareous rocks.

DISTRIBUTION: Widespread in western North America and the Great Lakes region (also on several other continents worldwide).

LITERATURE: Gaya, E. 2009. "Taxonomic Revision of the Caloplaca saxicola Group (Teloschistaceae, Lichen-Forming Ascomycota)." *Bibliotheca Lichenologica* 101: 1–191.

#4838 (PHOTO: J. LENDEMER)

Caloplaca sideritis

Iron Islands

Once learned, *Caloplaca sideritis* will be one of your most memorable lichen finds at White Rocks or elsewhere in the United States. When lacking apothecia, the species can be difficult to identify and might be mistaken for a *Verrucaria*, however unfortunate that is. But a careful search will generally yield discovery of the bright reddish-orange apothecia discs that characterize this species and make it unmistakable from anything else in our area. In addition, the apothecia are distinguished by having a proper margin that is brownish-black in color. Elsewhere in western North America, it is possible that *C. sideritis* could be confused with *C. pellodella* (a taxon that is rare in Colorado), but that species has a shiny thallus (versus dull in *C. sideritis*) and is generally squamulose at its margins (versus irregularly areolate throughout in *C. sideritis*). Check out Wetmore's (1996) very helpful publication on the "gray thallus group" of Caloplacas, and enjoy their subtleties.

CHEMISTRY: Thalloidima green and other anthroquinones; spot tests: K+ violet (apothecia margin), C+ violet (apothecia margin), KC+ violet (apothecia margin), P–, UV–.

SUBSTRATE AND ECOLOGY: On non-calcareous rocks and less commonly on calcareous rocks.

DISTRIBUTION: Primarily a midwestern US species, extending into southwestern North America; uncommon in eastern North America.

LITERATURE: Wetmore, C. M. 1996. "The *Caloplaca sideritis* Group in North and Central America." *Bryologist* 99: 292–314.

#4865 (PHOTO: J. LENDEMER)

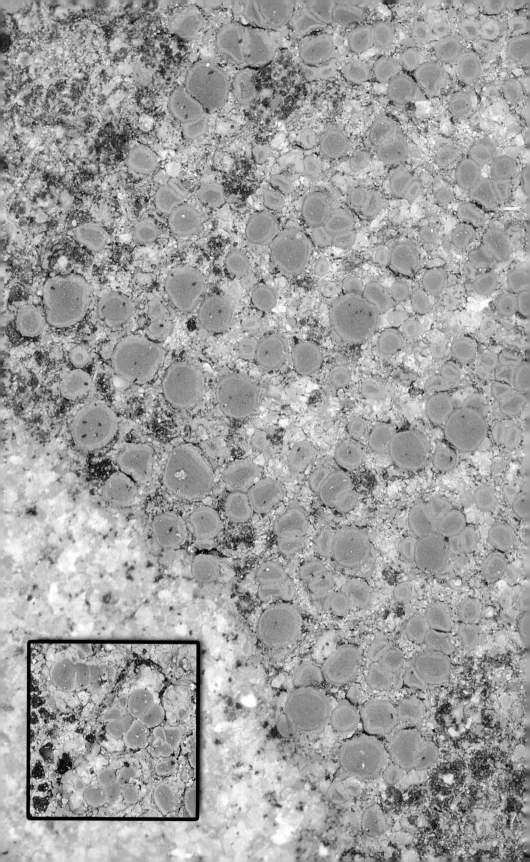

Caloplaca subsoluta

Orange Atoms

Caloplaca subsoluta will give you pause. At White Rocks, your life is made easy by the fact that there are only two species you could confuse it with: *C. atroflava* and *C. saxicola*. *Caloplaca subsoluta* differs from the former by having minute squamules (see inset) that are concolorous with the apothecia (versus a gray thallus) and differs from the latter by having only the most minute squamules (versus definitive lobes you could bet money on). In comparing *C. subsoluta* to other species in western North America, *Calopalca squamosa* is by far your biggest headache. The two species share a similar cell arrangement below the hypothecium (i.e., are paraplectenchymatous ... unfortunate word ... not my fault), but *C. squamosa* is definitely squamulose and has apothecia with thalline margins (versus apothecia without thalline margins in *C. subsoluta*).

CHEMISTRY: Emodin, fallacinal, parietin, teloschistin; spot tests: K+ red (epihymenium), C–, KC–, P–, UV–.

SUBSTRATE AND ECOLOGY: On non-calcareous or calcareous rocks. Why not just make a choice?

DISTRIBUTION: Widespread in North America wherever you find rocks.

LITERATURE: Wetmore, C. M. 2007. "Caloplaca." In T. H. Nash III, B. D. Ryan, C. Gries, and F. Bungartz, eds., *Lichen Flora of the Greater Sonoran Desert Region*, vol. 3. Tempe: Lichens Unlimited, Arizona State University, 179–220.

#4832 (PHOTO: J. LENDEMER)

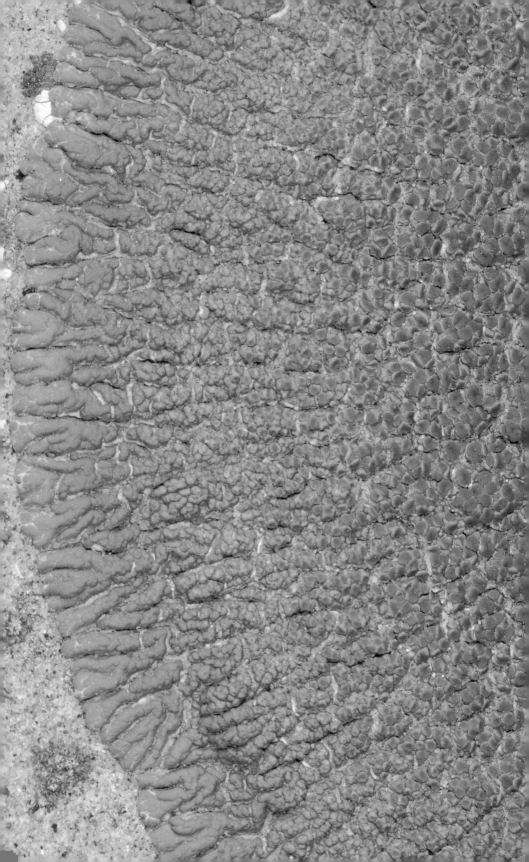

Caloplaca trachyphylla

Sunny Straps

This one should have been my prom date. *Caloplaca trachyphylla* is a large, spectacular, effigurate crustose lichen that at White Rocks is abundant on the large boulders atop the upper shelf of the sandstone formation. It is recognizable by its large, circular thalli that are abundantly fertile toward older (central) portions. At White Rocks, *C. trachyphylla* is most likely to be confused with a related saxicolous, orange lichen, *Xanthoria elegans*, but the latter differs by having slightly upturned lobes that are less tightly adnate to the substrate, a lower cortex (lacking in *C. trachyphylla*), and smaller thalli than those of *C. trachyphylla*. *Caloplaca trachyphylla* is also potentially confusable with other species of *Caloplaca* at White Rocks, but the latter consistently produce much smaller thalli with shorter lobes and do not resemble *C. trachyphylla* on any level other than spore type, thallus color, and chemistry. Recent phylogenetic analyses suggest that *C. trachyphylla* may more appropriately be treated in *Xanthomendoza* (Arup et al. 2013), but this transfer awaits deeper and more comprehensive phylogenetic sampling. *Caloplaca trachyphylla* is commonly associated with and presumably parasitized by *Acarospora stapfiana*. Look for both species throughout western North America.

CHEMISTRY: Emodin, fallacinal, parietin, parietinic acid, teloschistin; spot tests: K+ purple (cortex), C–, KC–, P–, UV–.

SUBSTRATE AND ECOLOGY: Commonly occurring on both calcareous and non-calcareous rocks.

DISTRIBUTION: Widespread in western North America at both high and low elevations, but lacking from the Pacific Northwest; also reported from several other countries worldwide.

LITERATURE: Arup, U., U. Sochting, and P. Froden. 2013. "A New Taxonomy of the Family Teloschistaceae." *Nordic Journal of Botany* 31: 16–83.

#4812 (PHOTO: E. TRIPP)

Candelariella clarkiae

Dina's Digression

Welcome to the city of Boulder—where you can still discover species new to science just a few miles from the University of Colorado. *Candelariella clarkiae* is a remarkable taxon and unmistakable among all other lichens at White Rocks because of its distinctive chartreuse color and crustose growth form with a weakly formed thallus of dispersed areoles (see inset). The distinctive thallus color and apothecia margins may not immediately suggest *Candelariella*, which is typically more egg-yolk yellow in color as in *C. rosulans* (pictured in the far right of main photo as well as overgrowing an apothecium of the new species in the inset). However, the presence of calycin and DNA sequence data clearly confirm its generic affinity. Besides color, *C. clarkiae* is marked by immersed, lecanorine apothecia with thick margins that are concolorous with the thallus and apothecia discs, which are green when young and mature to a pale salmon color. Elsewhere in our area, *C. clarkiae* might be confused for *C. vitellina* or *C. aurella*, from which it differs by not being polysporous (as in *C. vitellina*) or by occurring on silicaceous rocks and having a distinctive thallus (versus on calcareous rocks and lacking a distinctive thallus in *C. aurella*). See Westberg et al. 2011 (*Bryologist*) for keys to all North American *Candelariella*. This species was named to honor Dina Clark, collections manager at the COLO Herbarium, for her countless contributions to knowledge of Colorado botany. See further information on *C. clarkiae* under *Lecidea hoganii*.

CHEMISTRY: Calycin, pulvinic acid; spot tests: K–, C–, KC–, P–, UV+ dull orange.

SUBSTRATE AND ECOLOGY: Very rare on sandstone.

DISTRIBUTION: Currently known only from the type locality at White Rocks, Boulder, Colorado.

LITERATURE: Tripp, E. A., and J. C. Lendemer. 2015. "*Candelariella clarkiae* and *Lecidea hoganii*: Two Lichen Species New to Science from White Rocks Open Space, City of Boulder, Colorado." *Bryologist* 118: 154–63.

#4876 (PHOTO: E. TRIPP)

Candelariella rosulans

Over-Easy Lichen

Candelariella rosulans is, together with *Lecidella patavina*, one of the two most common lichens at White Rocks. Even though a highly variable species, it is easily recognized by its well-developed thallus of convex areoles or squamules that are egg-yolk yellow and matte in appearance. *Candelariella rosulans* plays an extremely important role ecologically, for it serves as one of the primary biotic soil crusts that facilitates "case hardening" of the loose, horizontal sandstones at White Rocks. This case hardening is further enhanced by montmorillinite clay present in sandstone. Together, the lichens and clay combine to slow the erosion of an already highly fragile substrate. Or, rather, do they speed the erosion of a highly fragile substrate by secreting substances that break down parent rock material? There is still so much to learn. Although an extremely widespread species in western North America, the type locality of *Candelariella rosulans* is just a few miles away: at ~6,000 feet elevation on Flagstaff Mountain. Boulder: be proud of your incredible lichen biota . . . Great Plains to great mountains!

CHEMISTRY: Calycin, vulpinic acid, pulvinic acid, pulvinic lactone; spot tests: K+ red (thallus), C–, KC–, P–, UV–.

SUBSTRATE AND ECOLOGY: Very common on silicaceous rocks such as sandstone. Rare on wood.

DISTRIBUTION: Pervasive in western North America at both low and high elevations, but lacking from humid regions of the Pacific Northwest.

LITERATURE: Readers who want to learn the *Candelariella* of the greater Front Range area or, more generally, of western North America should refer to the series of three publications by M. Westberg (2007a, 2007b, 2007c) and the follow-up treatment by Westberg et al. (2011).

#4877 (PHOTO: V. DÍAZ)

Dermatocarpon americanum

Lichen Lettuce

Dermatocarpon americanum is easily recognized to genus by its foliose growth form and umbilicate attachment. Its reproductive structures are that of perithecia rather than the more common apothecia, easily separating this genus from the somewhat superficially similar *Umbilicaria*, a genus not yet reported from White Rocks but one that is extremely common and widespread throughout Colorado, especially in montane ecosystems. *Dermatocarpon americanum* is characterized by its positive iodine reaction. In the past, this taxon has been confused with *D. miniatum*, but the latter species has an I- medulla. The only other species of *Dermatocarpon* at White Rocks is *D. moulinsii*, which is readily distinguished by its rhizinomorphs on the lower surface of the thallus (absent in *D. americanum*, although the lower surface of this species can be smooth to weakly rugose).

CHEMISTRY: No substances; spot tests: K–, C–, KC–, P–, I+ red (medulla), UV–.

SUBSTRATE AND ECOLOGY: Saxicolous on a variety of rock types.

DISTRIBUTION: Widespread in southwestern North America.

LITERATURE: Amtoft, A., F. Lutzoni, and J. Miadlikowska. 2008. "*Dermatocarpon* (Verrucariaceae) in the Ozark Highlands, North America." *Bryologist* 111: 1–40.

#4844 (PHOTO: J. LENDEMER)

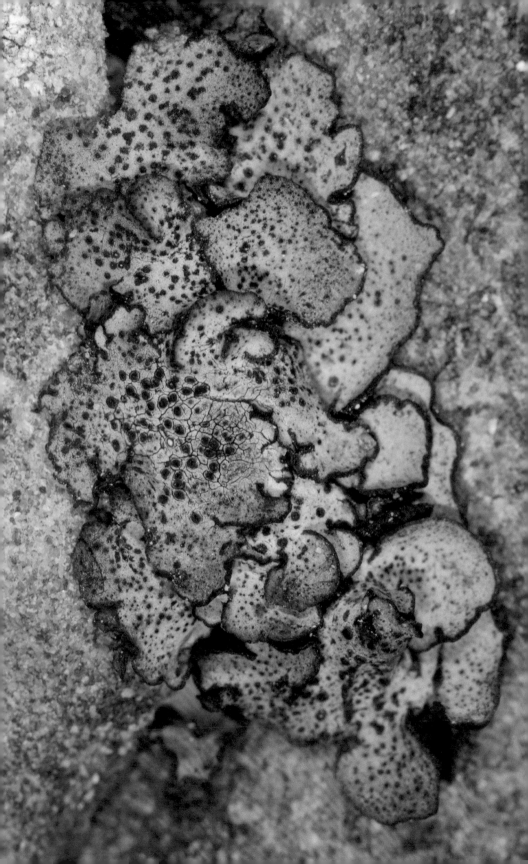

Dermatocarpon moulinsii

Stone Flakes

Like *Dermatocarpon americanum, D. moulinsii* is easily recognized to genus because of its foliose growth form with an umbilicate attachment to the substrate. Its perithecial reproductive structures easily distinguish the genus from the somewhat superficially similar genus *Umbilicaria. Dermatocarpon moulinsii* is readily distinguished from the only other species in the genus at White Rocks because of its rhizinomorphs on the lower surface. These structures are absent in *D. americanum,* although the lower surface of the latter species can be smooth to weakly rugose. When dry, which is often the case at White Rocks, species of *Dermatocarpon* have hard and noticeably crunchy thalli. Give it a little squish. It's like birding by ear.

CHEMISTRY: No substances; spot tests: K–, C–, KC–, P–, UV–.

SUBSTRATE AND ECOLOGY: Occasional on rocks, especially shaded rocks or in crevasses, as in the photo.

DISTRIBUTION: Relatively common in western North America and portions of the midwestern United States as well as the northern Appalachians (also reported from Europe and a few other areas).

LITERATURE: Amtoft, A., F. Lutzoni, and J. Miadlikowska. 2008. "*Dermatocarpon* (Verrucariaceae) in the Ozark Highlands, North America." *Bryologist* 111: 1–40.

#4828 (PHOTO: J. LENDEMER)

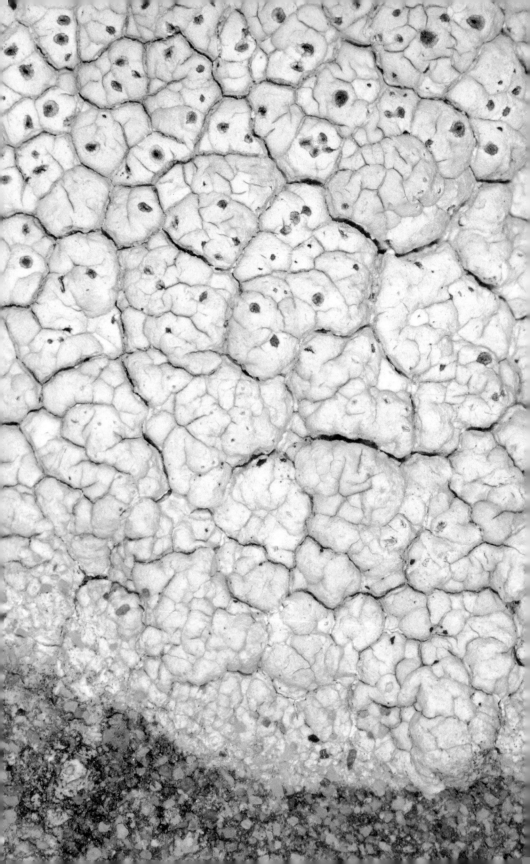

Diploschistes scruposus

Cowboy Float

Diploschistes scruposus is an extremely common and rather variable crustose species in Colorado and much of western North America that you will want to learn ASAP. It is most readily identifiable by its urceolate apothecia that give the appearance of immersed, crater-like fruiting bodies, its magnificent C+ red reaction, and its distinctive brown, muriform spores. Thalli of *D. scruposus* range from sordid white to gray in color. The species is most likely to be confused with other members of the genus, in particular *D. muscorum*. However, the latter has four spores per ascus that are much smaller in size (versus eight larger spores per ascus in *D. scruposus*) and also grows parasitically on other lichens. *Diploschistes scruposus* may overgrow other surfaces but is not parasitic. Elsewhere in Colorado, *D. scruposus* could also be confused with *D. diacapsis*, but that species is pruinose and mostly terricolous on dry, bare, arid soils.

CHEMISTRY: Diploschistesic acid, lecanoric acid, orsellinic acid; spot tests: K+ yellow or red (cortex) but sometimes K–, C+ red (cortex), KC+ red (cortex), P–, UV–.

SUBSTRATE AND ECOLOGY: Common on silicaceous rocks, but also frequently overgrowing mosses or other surfaces although not parasitic on these. Brodo et al. (2001) say that this species is "particularly aggressive, often overgrowing everything in its path." Another reason to stay active and not be sedentary for too long.

DISTRIBUTION: Extremely common throughout North American and Europe, also elsewhere worldwide, including Australia and South America.

LITERATURE: Brodo, I. M., S. D. Sharnoff, and S. Sharnoff. 2001. *Lichens of North America*. New Haven, CT: Yale University Press.

#4867 (PHOTO: E. TRIPP)

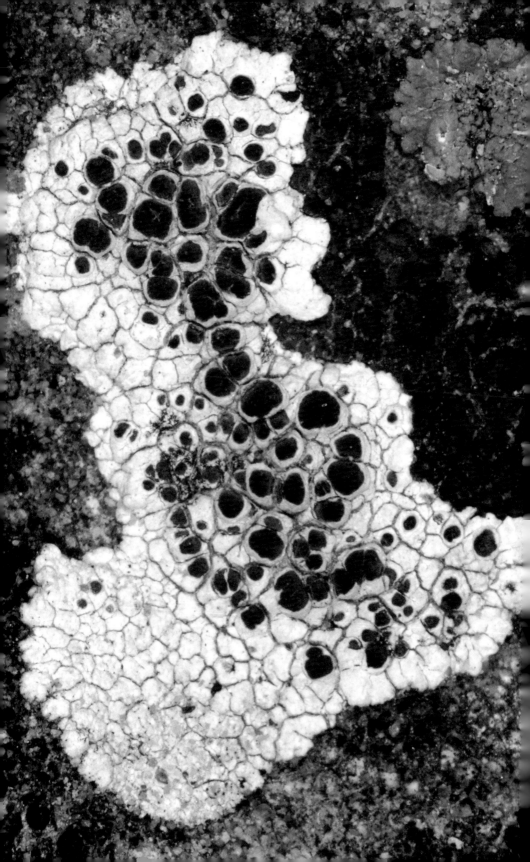

Diplotomma venustum

Doll's Eyes

Diplotomma venustum is a highly distinctive species that cannot be confused with any other at White Rocks because of its thick, chalky thallus with erupting black apothecia (often pruinose) and brown, three-septate spores. You might be tempted to call it *Lecidea hoganii*, but do a thin section of its apothecia with a razor blade, and note the major differences in spore morphology (hyaline and simple in the latter). This species is sometimes treated in the large, difficult, and in western North America important genus *Buellia*, but species with four-celled to muriform spores (versus two-celled) and thick, often lobed thalli are often placed in *Diplotomma*, and that taxonomy is followed here.

CHEMISTRY: No substances; spot tests: K–, C–, KC–, P–, UV–.

SUBSTRATE AND ECOLOGY: Saxicolous on a diversity of rock types, including sandstone, schist, calcareous rocks, and artificial substrates; occasionally parasitic on other lichens.

DISTRIBUTION: Great Plains and western North America, also in Europe.

LITERATURE: Bungartz, F., A. Nordin, and U. Grube. 2007 [also Nash 2008]. "Buellia." In T. H. Nash III, B. D. Ryan, C. Gries, and F. Bungartz, eds., *Lichen Flora of the Greater Sonoran Desert Region*, vol. 3. Tempe: Lichens Unlimited, Arizona State University, 113–79.

#4849 (PHOTO: J. LENDEMER)

Enchylium tenax

Little Jelly

Enchylium tenax: quite black, kind of hard-gooey. Feels unpleasant when you thin-section it with a razor blade, too. Alas, this one is in the elite club of special cyanolichens at White Rocks. The only two species it could possibly be confused with are *Lichinella stipatula* and *Lichinella nigritella*. Both of these are similarly dark black in color, but the former is dwarf fruticose and the latter has broader (usually > 2 mm wide versus usually < 2 mm wide in *E. tenax*) and more rounded lobes that ascend from the substrate. As with so many other lichens, the unfortunate truth is: it's better to learn the species than to learn the genus. Until recently, this species was treated in the larger and more widely recognized genus *Collema*, but phylogenetic and morphological analyses presented in Otálora et al. (2014) indicate the species is better treated in the genus *Enchylium*.

CHEMISTRY: No substances; spot tests: K–, C–, KC–, P–, UV–.

SUBSTRATE AND ECOLOGY: Soil or rock that is a bit soil-like, as in White Rocks.

DISTRIBUTION: Widespread in western North America and the upper Midwest.

LITERATURE: Otálora, M., P. Jørgensen, and M. Wedin. 2014. "A Revised Generic Classification of the Jelly Lichens, Collemataceae." *Fungal Diversity* 64: 275–93.

#4878 (PHOTO: V. DÍAZ)

Endocarpon pallidulum

Western Crepes

Endocarpon pallidulum is a distinctive species by its brown, squamulose to almost scaly thallus with perithecia that have darkened walls, are sunken within each scale, and are visible as emergent black dots toward the center of scales. The genus as a whole is distinctive in perithecial cross section by having muriform spores and algal cells contained within the hymenium. Following sectioning, *E. pallidulum* can only be confused with *Staurothele* at White Rocks, which also has brown, muriform spores and algae in the hymenium. However, *Staurothele* differs by having squamules that arise individually from a prothallus rather than an areolate thallus derived from cracking of a continuous thallus. In addition, *Endocarpon* has a clearly differentiated lower cortex often with hyphae or rhizine-like attachment structures, whereas *Staurothele* lacks a lower cortex and attachment structures (note: when highly congested, the scales of *E. pallidulum* can give the thallus an almost continuous, scum-like appearance; look for a lower cortex). *Endocarpon pallidulum* can be differentiated from other species in the genus by the number of spores per ascus (two/ascus), the presence of small rhizohyphae on the lower cortex (versus larger rhizines), and its lower surface color. The upper surface is highly variable in color depending on its hydration state. When wet, it is bright green (left inset); when semi-hydrated, it is brown (main photo); when dessicated, it is brownish-gray (right inset), just like the rest of us.

CHEMISTRY: No substances; spot tests: K–, C–, KC–, P–, UV–.

SUBSTRATE AND ECOLOGY: Relatively common on calcareous and non-calcareous rock and on soils over rocks.

DISTRIBUTION: Widespread in the Great Plains and portions of eastern North America, as well as the southwestern United States; also several other areas worldwide.

LITERATURE: Breuss, O. 2007a. "Endocarpon." In T. H. Nash III, B. D. Ryan, C. Gries, and F. Bungartz, eds., *Lichen Flora of the Greater Sonoran Desert Region*, vol. 1. Tempe: Lichens Unlimited, Arizona State University, 181–87.

#4850, 4860, 4862 (PHOTOS: E. TRIPP)

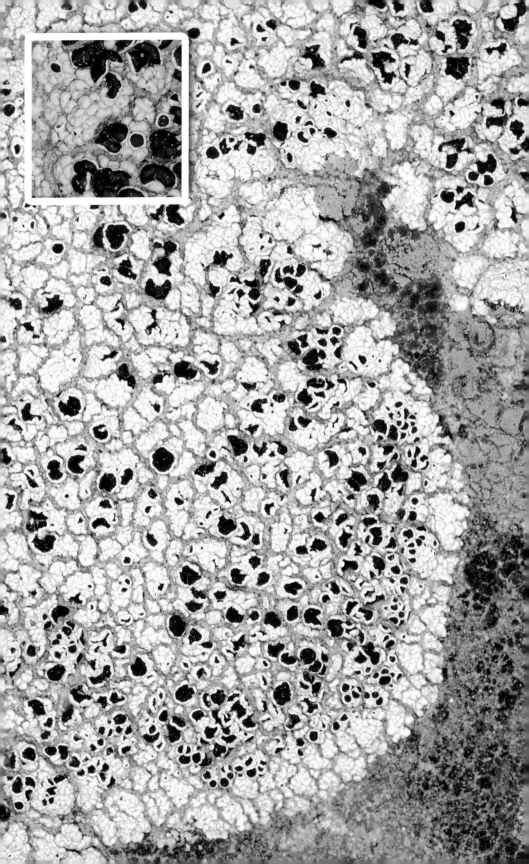

Lecanora argopholis

Butter Biscuits

Lecanora argopholis is a verrucose to areolate lichen that produces proper, albeit very short, lobes (0.5–1.5 mm long) at its margins. At first, it may be confused for other placidioid Lecanoras in our area (e.g., *L. garovaglii*, *L. novomexicana*, *L. muralis*), but once learned it cannot be mistaken for any of these. In addition to its characteristic short lobes, *L. argopholis* has a K+ yellow thallus and apothecia discs that are always reddish-brown in color and epruinose. The thallus color of *L. argopholis* ranges from gray to yellow, which is attributable to varying levels of the yellow xanthone epanorin. However, morphology in combination with thin layer chromatography, which must be conducted for all placidioid Lecanoras, will readily facilitate identification of the species. Most thalli of *L. argopholis* at White Rocks are pale greenish-yellow, as in figure at left. *Lecanora* is the most species-rich genus of lichens in North America and this is reflected at White Rocks, where five different species (the most of any one genus there) co-occur. Are they closely related? If so, how are they reproductively isolated?

CHEMISTRY: Atranorin, epanorin, zeorin, fatty acids; spot tests: K+ yellow (cortex), C–, KC+ yellow (cortex), P+ yellow (cortex) but sometimes P–.

SUBSTRATE AND ECOLOGY: Common on calcareous and silicaceous rocks, sometimes on mosses or plant debris, often found in exposed sites. This species has a broad ecological niche, including arctic rock outcroppings down to prairie environments such as found at White Rocks.

DISTRIBUTION: Widespread in mountainous areas of western North America (except for the Sierra Mountains), the northern Great Plains, and the Great Lakes region (also in Europe, Africa, and Asia).

LITERATURE: Brodo, I. M., S. D. Sharnoff, and S. Sharnoff. 2001. *Lichens of North America*. New Haven, CT: Yale University Press.

#4819 (PHOTO: E. TRIPP)

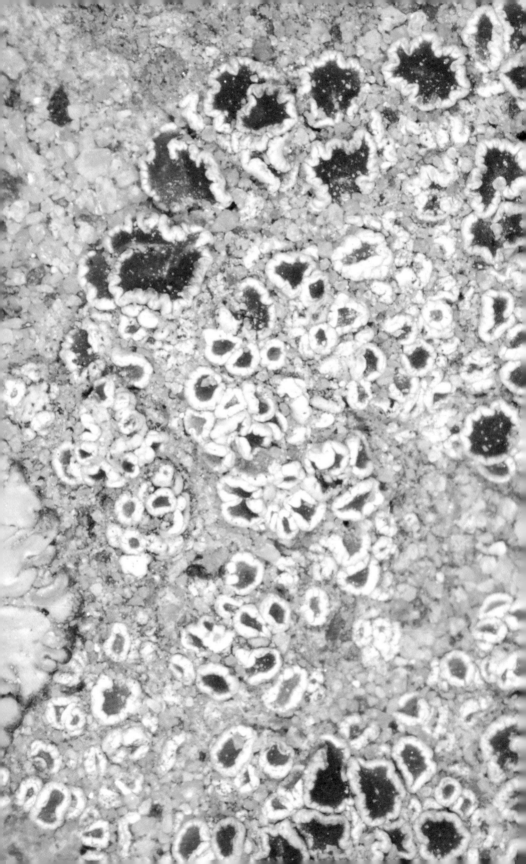

Lecanora flowersiana

Seville's Lecanora

Lecanora flowersiana is one of several difficult-to-identify species in the *Lecanora dispersa* group, which lack visible thalli. At White Rocks, *L. flowersiana* is most likely to be confused with a new, unnamed species of *Lecanora*, included in this Field Guide as *Lecanora* sp. nov. The latter also lacks an obvious thallus but differs by the presence of usnic acid, among other features. Elsewhere, *L. flowersiana* is most likely to be confused with *L. crenulata* or *L. hagenii*, from which it differs by having epruinose discs and larger spores (versus heavily pruinose and smaller spores in *L. crenulata*) and much thicker apothecial margins (versus very thin in *L. hagenii*). Of other species that lack granules in the hymenium, *L. invadens* and *L. percrenata* occur on calcareous rock. Apothecial discs of *L. flowersiana* range from reddish-brown to dark brown to black. Readers who want to learn more about this challenging group should refer to Śliwa's (2007) monograph for further information.

CHEMISTRY: No substances; spot tests: K–, C–, KC–, P–, UV–.

SUBSTRATE AND ECOLOGY: Occasional on dry, exposed sandstone and granite; rare on wood.

DISTRIBUTION: Restricted to temperate regions of central and western North America.

LITERATURE: Śliwa, L. 2007. "A Revision of the *Lecanora dispersa* Complex in North America." *Polish Botanical Journal* 52: 1–70.

#4815 (PHOTO: E. TRIPP)

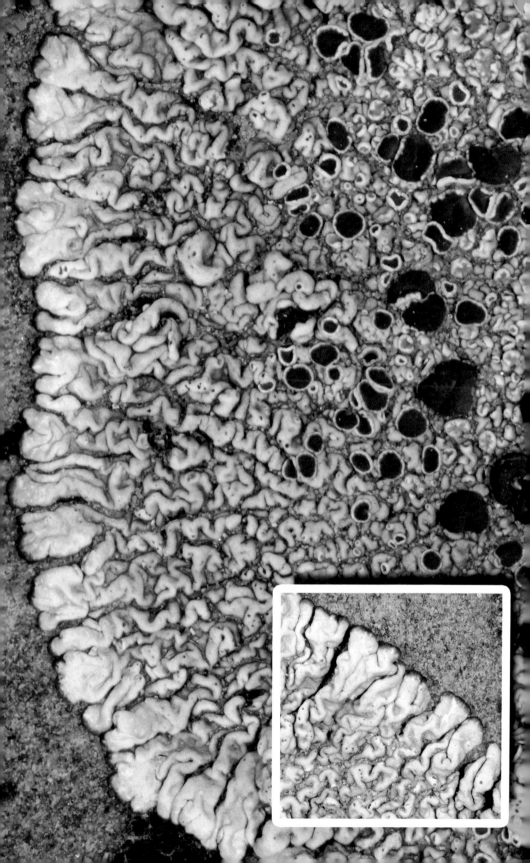

Lecanora garovaglii

A Textured Dilemma

Lecanora garovaglii is, together with *L. argopholis* and *L. muralis*, one of three placidioid Lecanoras at White Rocks. It is easily distinguished from both by its elongated lobes (versus very short in *L. argopholis*) that are sinuous to folded near the margins, thus appearing somewhat swollen (versus lobes very short and not sinuous or swollen in *L. muralis*). *Lecanora garovaglii* is most likely to be confused with another placidioid species not yet found at White Rocks but one that is present and indeed common in Colorado (especially at higher elevations and in less arid habitats): *L. novomexicana*. From the latter, *L. garovaglii* can be distinguished by thin layer chromatography (zeorin versus psoromic acid present in the medulla of *L. novomexicana*), the C– medulla reaction (versus C+ red in *L. novomexicana*), and spongier medulla tissue (versus more solid in *L. novomexicana*). Apothecial discs of *L. garovaglii* range in color from yellow to brown to bluish-black.

CHEMISTRY: Usnic acid, placodiolic and/or isousnic acid, zeorin (medulla) unknown terpenoids (medulla); spot tests: K– or K+ pale yellow (medulla), C–, KC+ gold (cortex), P–, UV–.

SUBSTRATE AND ECOLOGY: Common on saxicolous surfaces, especially sandstone in arid areas. The species is not well represented in Boulder County.

DISTRIBUTION: A specialty of much of western North America.

LITERATURE: Brodo, I. M., S. D. Sharnoff, and S. Sharnoff. 2001. *Lichens of North America*. New Haven, CT: Yale University Press.

#4813 (PHOTO: E. TRIPP)

Lecanora muralis

My Old Friend

Lecanora muralis is one three placidioid Lecanoras at White Rocks and one of the most common saxicolous lichens in the city of Boulder (and indeed, much of the West). It is easily differentiated from both *L. argopholis* and *L. garovaglii* by its flattened morphology in which the entire thallus is closely appressed to the substrate (thalli of *L. argopholis* and *L. garovaglii* are more three-dimensional) and its tendency to form small, tidy rosettes. In addition, the presence of leucotylin as determined by thin layer chromatography is distinctive for this species. *Lecanora muralis* is always an awkward shade of greenish-yellow, just like the photo, because of the presence of usnic acid in the cortex. From the inset photo, note the darkened lobe tips and often pale (but not pruinose) lobe margins, typical of the species. *Lecanora muralis* is an extremely wide-ranging and variable species that has been the subject of much taxonomic splitting. It is possible that, following future studies that integrate morphology, chemistry, molecular work, and magic, a different name may be applied to our specimens. Make this species the first you learn and love among placidioid Lecanoras. Besides, if you tilt your head the right way, it's already shaped like a heart.

CHEMISTRY: Leucotylin, usnic acid, zeorin, triterpenes, unknown fatty acids, sometimes with isousnic acid; spot tests: K–, C–, KC+ yellow (cortex), P–, UV–.

SUBSTRATE AND ECOLOGY: Very common on most rock types, including artificial urban surfaces such as rock walls. Look for this species covering the north-facing rock wall that borders the north side of the University of Colorado's Macky Auditorium.

DISTRIBUTION: Widespread in the United States but conspicuously absent from the Southeast; also present in southern Canada, northern Mexico, Europe, and on most other continents worldwide.

LITERATURE: Ryan, B. D., H. T. Lumbsch, M. I. Messuti, C. Printzen, L. Sliwa, and T. H. Nash III. 2007. "*Lecanora*." In T. H. Nash III, B. D. Ryan, C. Gries, and F. Bungartz, eds., *Lichen Flora of the Greater Sonoran Desert Region*, vol. 2. Tempe: Lichens Unlimited, Arizona State University, 176–286.

#4822 (PHOTO: E. TRIPP)

Lecanora sp. nov.

Crack Pots

Lecanora sp. nov. is one of several challenging species in the *Lecanora dispersa* complex. Species in this group have thalli that are immersed entirely within the substrate (rock or bark) and differ subtly in features such as spore size and shape, apothecia morphology and anatomy, and ecology. In Colorado, we have ~ten species in the complex, and the best way to distinguish them is by use of Śliwa's (2007) keys. This new species is still under investigation but can be distinguished from other members of the complex in our area by the presence of usnic acid in combination with its cracked apothecia margins and birefringent crystals (visible using polarized light) in the hypothecium. It is perhaps most likely to be confused with *Lecanora dispersa*, from which it differs most obviously by the usnic acid and by lack of a P+ orange reaction in the apothecial margins.

CHEMISTRY: Usnic acid; spot tests: K–, C–, KC–, P– but sometimes P+ orange (apothecia margin), UV–.

SUBSTRATE AND ECOLOGY: So far as known, this species is restricted to sandstone subtrates.

DISTRIBUTION: This species is known only from two collections in Colorado: one at White Rocks and one from a similar elevation on a sandstone formation in Montrose County. Recently, though, I found a very large population of this species at the smaller white rocks formation/outcropping in Niwot. This site isn't far from the main White Rocks reported on in this Field Guide, but it represents another and very important additional population of this species.

LITERATURE: None (yet).

#4823 (PHOTO: E. TRIPP)

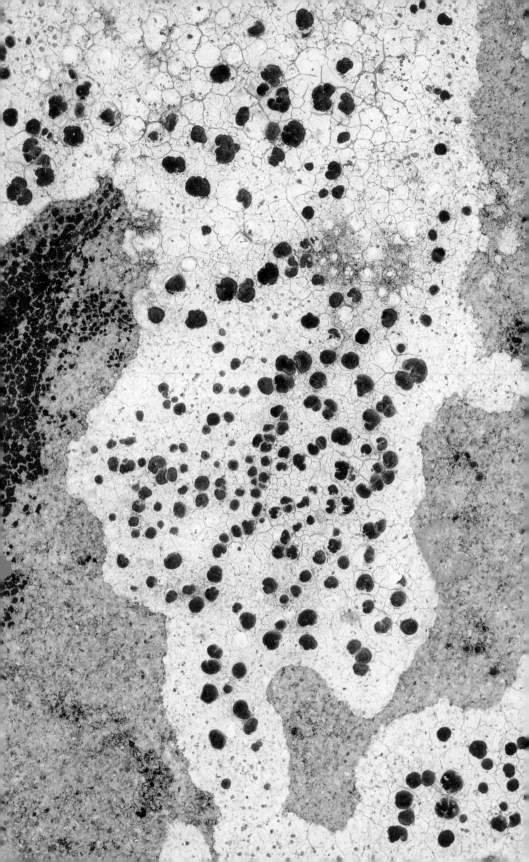

Lecidea hoganii

Timscape

Brace yourself, or this species will woo you right off your feet. *Lecidea hoganii* is a true charmer. At White Rocks, this species cannot be confused with any other lichen because of its thick (up to 0.6 mm), chalky white thallus with black apothecia that are sessile to raised on columns of concolorous thalline tissue, its dark brownish–black epihymenium, its hymenium that is variably stained with streaks of an unknown pink pigment, and its lack of chemistry. The most superficially similar species at White Rocks is *Diplotomma venustum*, but that species has brown, multi-septate spores (versus hyaline, simple spores in *L. hoganii*). Both *Lecidella carpathica* and *Lecidella patavina* have hyaline, simple spores but differ morphologically in many ways, including their blue-green epihymenia. Outside of White Rocks, *L. hoganii* might be confused with *L. saximontana* ined., but the latter also has a blue-green epihymenium as well as a thin, poorly developed thallus. *Lecidea* is one of the most difficult genera on the planet—it's true—but that also makes it one of the most exciting. *Lecidea hoganii* was the first lichen I collected at White Rocks, and *Candelariella clarkiae* was the last. The Primary and Recency Effect—looks like the psychologists had it right all along. This species was named in honor of Tim Hogan, collections manager at the COLO Herbarium, for his lifetime of contributions to knowledge of the Colorado flora.

CHEMISTRY: No substances; spot tests: K–, C–, KC–, P–, UV–.

SUBSTRATE AND ECOLOGY: Known only from Fox Hills Formation sandstone.

DISTRIBUTION: Currently known only from the type locality at White Rocks, Boulder, Colorado.

LITERATURE: Tripp, E. A., and J. C. Lendemer. 2015. "*Candelariella clarkiae* and *Lecidea hoganii*: Two Lichen Species New to Science from White Rocks Open Space, City of Boulder, Colorado." *Bryologist* 118: 154–63.

#4805 (PHOTO: E. TRIPP)

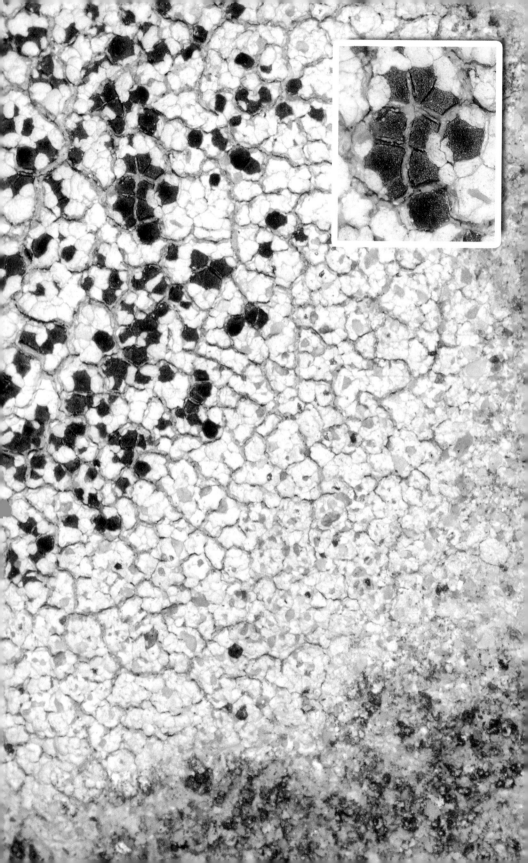

Lecidea tessellata

Hornswoggle Lichen

Lecidea tessellata is one of the most common saxicolous lichens in North America, known from hundreds if not thousands of collections in the western United States alone. It is characterized by its sunken, black, angular apothecia (see inset) against a sordid white, dull thallus of somewhat regularly shaped areoles (see main photo). The species also has an I+ blue medulla and a conspicuous black prothallus, but the latter is lacking among White Rocks specimens (see below). *Lecidea tessellata* might be confused with several other taxa, one of which is *Lecanora oreinoides*. The latter differs from *L. tessellata* by having a distinctively shiny cortex and is primarily a southeastern US taxon. *Lecidea* and close relatives are desperately in need of taxonomic revision. At present, *L. tessellata* is the best-fitting name for the taxon at White Rocks, but following proper and thorough revision of the genus (not by me), our White Rocks species as well as others in our area currently attributed to *Lecidea tessellata* will mostly likely be called by a different name. For starters, perhaps you can observe the slightly bluish hue of the thallus of this collection, which is somewhat typical of other *Lecidea tessellata* in our area. *Lecidea* is arguably the most difficult genus of lichens in western North America.

CHEMISTRY: Confluentic acid; spot tests: K–, C–, KC–, P–, I+ blue (medulla), UV–.

SUBSTRATE AND ECOLOGY: Common on non-calcareous rock types, primarily.

DISTRIBUTION: Extremely widespread in the western United States, occasional in the Great Plains and also in the Appalachians where it reaches its southeastern distributional limit in the mountains of western North Carolina; known elsewhere in the world, primarily northern Canada and Europe.

LITERATURE: Brodo, I. M., S. D. Sharnoff, and S. Sharnoff. 2001. *Lichens of North America*. New Haven, CT: Yale University Press.

#4865 (PHOTO: E. TRIPP)

Lecidella carpathica

Lichen Moguls

Lecidella carpathica is one of several difficult crustose lichens in the *Lecidea /
Lecidella* group at White Rocks. Its saxicolous habit, areolate to bullate,
thickened white thallus, bluish-green exciple, yellowish-brown hypothecium,
and chemistry (determined by thin layer chromatography) help identify this
lichen to species. *Lecidella carpathica* is most likely to be confused initially
for a *Lecidea*, *Buellia*, or *Rhizocarpon*, but the latter two genera are characterized by species with brown, transversely septate spores. *Lecidella* is separated
from *Lecidea* by technical aspects of ascus anatomy. See further comments
under *Lecidea tessellata*.

CHEMISTRY: Atranorin; spot tests: K+ yellow (cortex), C–, KC–, P+ yellow
(cortex), UV–.

SUBSTRATE AND ECOLOGY: Saxicolous on non-calcareous rocks; occasionally
lignicolous or corticolous.

DISTRIBUTION: Widespread in western North America (especially abundant
throughout Boulder County) and the upper Great Lakes region, also in
northern Canada, northern Europe, and Siberia.

LITERATURE: Knoph, J.-G., and C. Leuckert. 2007. "Lecidella." In T. H. Nash
III, B. D. Ryan, C. Gries, and F. Bungartz, eds., *Lichen Flora of the Greater
Sonoran Desert Region*, vol. 2. Tempe: Lichens Unlimited, Arizona State
University, 309–20.

#4875 (PHOTO: V. DÍAZ)

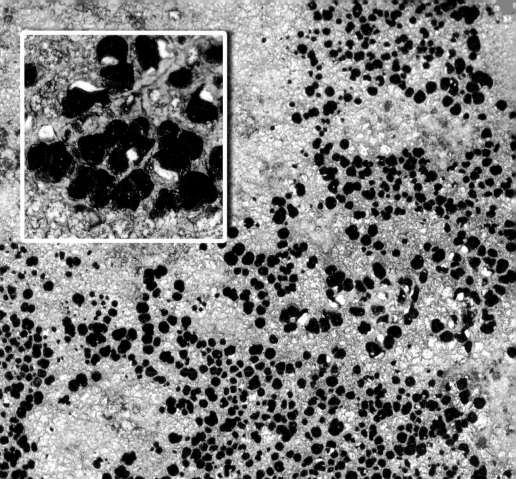

Lecidella patavina

Snail Snacks

Lecidella patavina is, together with *Candelariella rosulans*, one of the two most common lichens at White Rocks. The most noticeable part of this lichen is its very dark lecideine apothecia that lack algae in the margins. The thallus of *L. patavina* is thin but well developed and consists of continuous to irregular, sordid white to beige areolae (in figure at left, the thallus more or less blends in with the Fox Hills Formation sandstone). *Lecidella patavina* is identifiable to species by its saxicolous habit, bluish-green exciple, chemistry, and hymenium interspersed with oil droplets. Damaged apothecia (see inset) are common at White Rocks and may be the result of snail or other invertebrate feed activity, and although little empirical data documents the importance of lichens as a food resource for other organisms (but see Cameron 2009; Newmaster et al. 2013), surely these must be tasty to something. See text under other *Lecidea* and *Lecidella* for additional taxonomic information.

CHEMISTRY: Atranorin, diploicin (hymenium), lichexanthone; spot tests: K– or K+ yellow (cortex), C– or C+ yellowish-red (cortex), KC– or KC+ yellowish-red (cortex), P– or P+ yellow (cortex), UV–.

SUBSTRATE AND ECOLOGY: As currently understood, common on both calcareous and non-calcareous rocks.

DISTRIBUTION: A common saxicolous species from the southern Rockies to the Arctic and adjacent High Plains; also in northern Europe and reported from Siberia.

LITERATURE: Knoph, J.-G. 1990. "Untersuchungen an gesteinsbewohnenden xanthonhaltigen Sippen der Flechtengattung *Lecidella* (Lecanoraceae, Lecanorales) unter besonderer Berücksichtigung von außereuropäischen Proben exklusive Amerika." *Bibliotheca Lichenologica* 36: 1–183.

#4869 (PHOTO: E. TRIPP)

Lecidella stigmatea

Silver Fox

Lecidella stigmatea is one of five species in the *Lecidea* / *Lecidella* group with white or grayish-white thalli at White Rocks. It is characterized by its strongly bullate to areolate, white to gray thallus, its bluish-green epihymenium, its non–oil-inspersed hymenium, its hyaline hypothecium, and its chemistry. *Lecidea hoganii* has a much less bullate thallus and a brownish-black epihymenium, and it lacks secondary compounds. *Lecidella patavina* has an oil-inspersed hymenium and generally a more poorly developed thallus. *Lecidella carpathica* has a similar thallus but has a yellowish-brown hypothecium and the hymenium contains diploicin, unlike *L. stigmatea*. Finally, *Lecidea tessellata* differs by its angular apothecia, confluentic acid, and an I+ blue medulla. *Lecidea* and *Lecidella* need a massive taxonomic revision. Be the one.

CHEMISTRY: Atranorin, zeorin, lichexanthone; spot tests: K+ yellow (cortex), C–, KC–, P–, UV–.

SUBSTRATE AND ECOLOGY: On both calcareous and non-calcareous rocks. A taxonomic revision might change things.

DISTRIBUTION: Widespread in North America except for the Southeast. A taxonomic revision might change things.

LITERATURE: Brodo, I. M., S. D. Sharnoff, and S. Sharnoff. 2001. *Lichens of North America*. New Haven, CT: Yale University Press.

#4846 (PHOTO: J. LENDEMER)

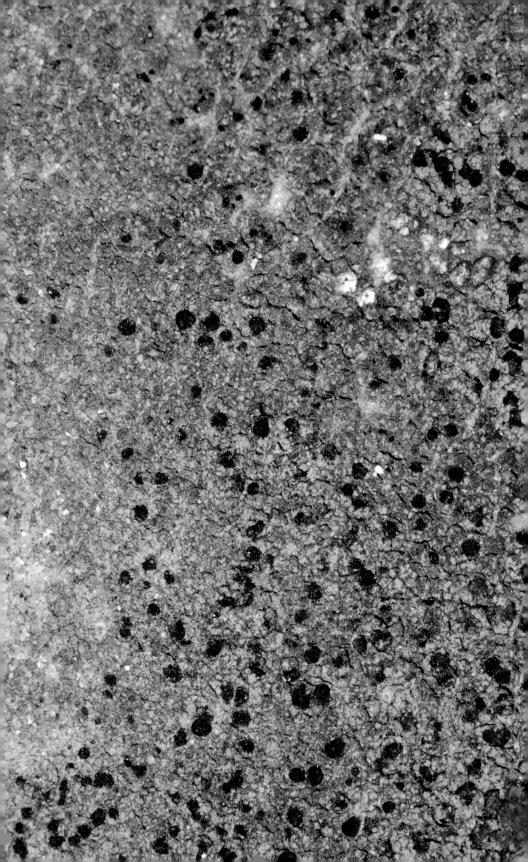

Lecidella viridans

Eggplant in Garlic Sauce

You got this one! Black dots, lecideine apothecia, simple hyaline spores. Must be a *Lecidea* or *Lecidella*, right? Right! *Lecidella viridans* is the only species in this group at White Rocks that doesn't have a distinctly white or grayish-white thallus. Instead, the thallus of *L. viridans* is composed of army-green to dull yellow granules, as the specific epithet suggests. Its black apothecia are the typical sort for this group, whose characters do not warrant repetition. Go with your instinct. It's probably easier to cope with than is the chemistry of this species.

CHEMISTRY: 4,5-dicloronorlichexanthone, arthothelin, isoarthothelin, thiophanic acid; spot tests: K–, C+ dull red (cortex), KC+ dull red (cortex), P–, UV–.

SUBSTRATE AND ECOLOGY: On non-calcareous rocks, so far as understood.

DISTRIBUTION: Primarily a species of the southwestern United States, but also reported from the Southeast.

LITERATURE: Knoph, J.-G., and C. Leuckert. 2007. "Lecidella." In T. H. Nash III, B. D. Ryan, C. Gries, and F. Bungartz, eds., *Lichen Flora of the Greater Sonoran Desert Region*, vol. 2. Tempe: Lichens Unlimited, Arizona State University, 309–20.

#4837 (PHOTO: J. LENDEMER)

Lepraria finkii

Spring Confetti

Lepraria finkii is the only member of the genus present at White Rocks and is unmistakable among all other lichens there because of its morphology and ecology. As is typical for the genus, *L. finkii* is composed entirely of small, asexual, ecorticate granules that become overlapping with time, resulting in a thick, soft, and poofy crust over rock surfaces. *Lepraria finkii* is the most common and widely distributed member of the genus in North America and the most common species of *Lepraria* in the Front Range of the southern Rocky Mountains. Its ubiquity is in part related to its broad ecological tolerance, occurring in a wide variety of habitats as long as shade and some moderate to high relative humidity can be found. Indeed, at White Rocks it occurs only under the deep rock overhang that characterizes most of the lower tier of the formation, just above Boulder Creek—an area that stays shaded, cool, and moist year-round, qualities many of us strive to achieve. Thin layer chromatography is essential to correctly identify this taxon to species, as it is for almost all others in the genus.

CHEMISTRY: Atranorin, stictic acid, zeorin, and traces of roccellic/angardianic acid, norstictic acid, stictic acid aggregate, and two unknowns; spot tests: K+ yellow-brown (granules), C−, KC+ yellow-brown (granules), P+ orange (granules), UV−.

SUBSTRATE AND ECOLOGY: Very common on saxicolous substrates of all types (it is the only member of the genus that colonizes calcareous rocks in addition to silicaceous ones), occasionally migrating onto mosses, soil, other lichens, or you if you stand still long enough. *Lepraria finkii* has a remarkable tolerance to highly polluted areas, including most urban and developed areas of the Front Range. It occurs in every borough of New York City.

DISTRIBUTION: Widespread in North America (less common in the West than the East); also in Europe.

LITERATURE: Lendemer, J. C. 2013. "A Monograph of the Crustose Members of the Genus *Lepraria* Ach. s. str. (Stereocaulaceae, Lichenized Ascomycetes in North America North of Mexico)." *Opuscula Philolichenum* 11: 27–141.

#4825 (PHOTO: E. TRIPP)

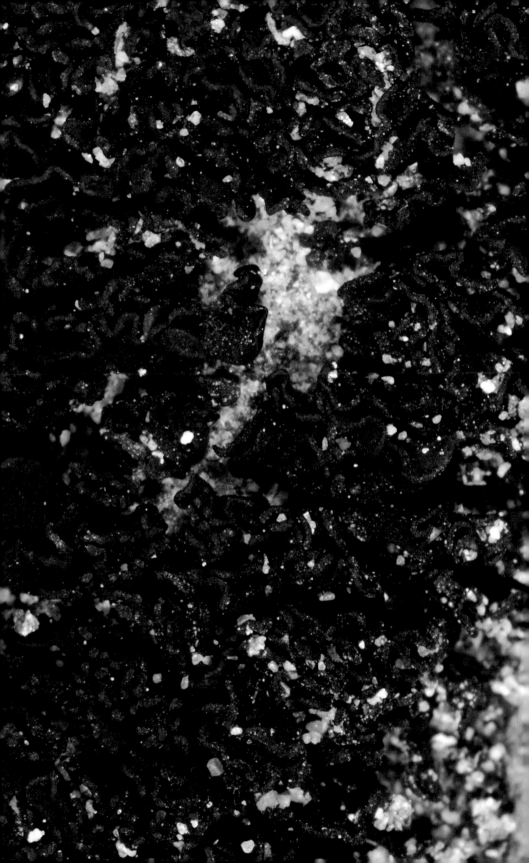

Lichinella nigritella

Volcanic Kelp

This is a fun find in the field. Lichen? Cyanobacterium? Free-living fungus? Definitely the first! *Lichinella nigritella* is extremely well differentiated from the only other congener at White Rocks, *L. stipatula*, because of its foliose growth form (versus dwarf fruticose and composed of tiny cylindrical lobes in *L. stipatula*). *Lichinella nigritella* has broadly rounded lobes that are distinctly ascending from the substrate. No sexual reproductive structures (i.e., thallinocarps in this genus) were observed at White Rocks.

CHEMISTRY: No substances; spot tests: K–, C–, KC–, P–, UV–.

SUBSTRATE AND ECOLOGY: Common on suitable saxicolous substrates, which includes both siliceous and calcareous rocks.

DISTRIBUTION: Relatively common throughout the southwestern United States, as well as in suitable habitats in the southern Rockies and the Great Plains. Elsewhere in our area, look for this species growing on Lyons Sandstone in the Hall Ranch area.

LITERATURE: Schultz, M. 2007. "Lichinella." In T. H. Nash III, B. D. Ryan, C. Gries, and F. Bungartz, eds., *Lichen Flora of the Greater Sonoran Desert Region*, vol. 3. Tempe: Lichens Unlimited, Arizona State University, 233–42.

#4807 (PHOTO: E. TRIPP)

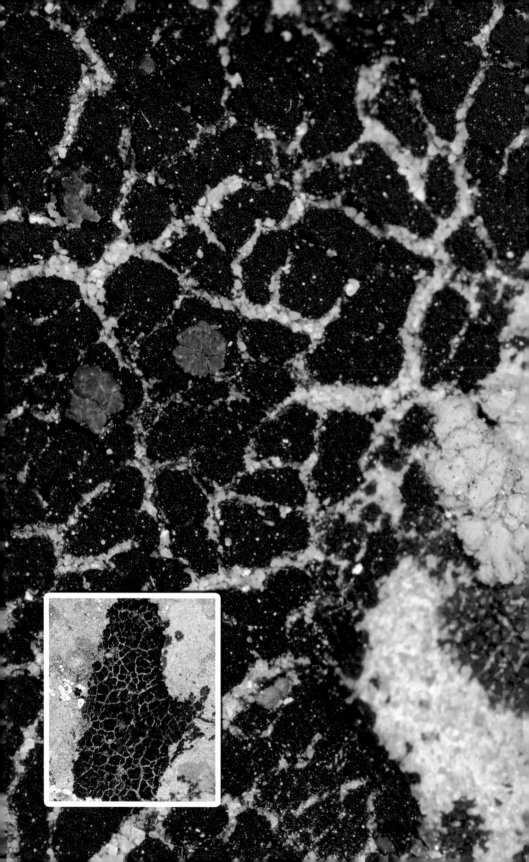

Lichinella stipatula

Black Forest

Lichinella stipatula may at first be mistaken as a cyanobacterium or otherwise overlooked as something not lichenological. In fact, it is one of only three cyanolichens at White Rocks (the others being *Enchylium tenax* and *L. nigritella*). *Lichinella stipatula* is a dwarf fruticose is a dwarf fruticose species composed of tiny cylindrical lobes, thus giving an overall appearance of minute black cushions and a micro-maze for some collembola to wander through. This morphology combined with its very black thallus color readily distinguishes it from any other lichen at White Rocks. *Lichinella stipatula* is the most common member of the genus in western North America. Although commonly found with fruiting bodies, none were seen at White Rocks.

CHEMISTRY: No substances; spot tests: K–, C–, KC–, P–, UV– (exciting).

SUBSTRATE AND ECOLOGY: Locally common in suitable habitats where it occupies silicaceous rocks.

DISTRIBUTION: Occasional in the western United States (especially the Southwest) and Great Plains; also reported from several other regions worldwide.

LITERATURE: Schultz, M. 2005. "An Overview of *Lichinella* in the Southwestern United States and Northwestern Mexico, and the New Species *Lichinella granulosa*." *Bryologist* 108: 567–90.

#4803B (PHOTO: E. TRIPP)

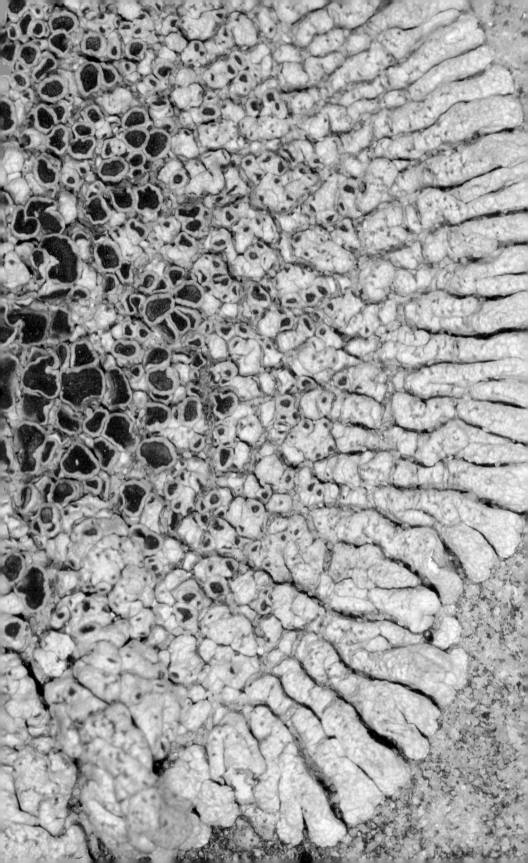

Lobothallia alphoplaca

Backcountry Pancakes

Lobothallia alphoplaca is at first glance superficially similar to some placidi-oid Lecanoras at White Rocks. However, the species is easily distinguished from all three (*Lecanora argopholis, L. garovaglii, L. muralis*) by its sordid white to gray to pale brown thallus color and K+ red cortex reaction (the three Lecanoras are greenish-yellow because of the presence of usnic acid or atranorin [both of which are lacking in *Lobothallia alphoplaca*] and do not produce a K+ red cortex reaction). Indeed, no other squamulose lichen at White Rocks has the thallus coloration and characteristically sunken young apothecia (becoming emergent in the oldest fruiting bodies) typical of *Lobothallia*. Once learned, the genus is unmistakable in the western United States. Species of *Lobothallia* were previously treated within *Aspicilia*, with which they share small, simple, hyaline spores and lecanorine apothecia margins, but later authors segregated species with distinctly lobed thalli into *Lobothallia* (convenient name), and indeed those taxa form a monophy-letic group (Nordin et al. 2010). *Lobothallia alphoplaca* is highly variable in thallus color, ranging from brown to white. Its convex to nearly terete lobes therefore best differentiate it from *Lobothallia praeradiosa*, which has less convex lobes. The latter species is not known from White Rocks but does occur in the area.

CHEMISTRY: Norstictic, constictic, or salazinic acid; spot tests: K+ red (cortex), C–, KC–, P+ orange (medulla), UV–.

SUBSTRATE AND ECOLOGY: Common on silicaceous rocks, especially in exposed areas.

DISTRIBUTION: Throughout western North America; also known from several other continents.

LITERATURE: Nordin, A., S. Savic, and L. Tibell. 2010. "Phylogeny and Taxonomy of *Aspicilia* and Megasporaceae." *Mycologia* 102: 1339–49.

#4832 (PHOTO: J. LENDEMER)

Montanelia disjuncta

Creeping Olivine

Montanelia disjuncta is one of two species in the genus (formerly *Melanelia*) found at White Rocks. Both *M. disjuncta* and *M. tominii* are easily recognized by their adnate to loosely adnate attachment to the substrate, their dark olive to brown thalli with pseudocyphellae (cracks in upper cortex exposing the medulla) best seen near the lobe tips, and their laminal or submarginal soralia that give rise to soredia. Both species occupy similar environments and have similar worldwide distributions. *Montanelia disjuncta* can, however, easily be separated from *M. tominii* by its C− medulla reaction (versus C+ bright red in *M. tominii*). In addition, pseudocyphellae tend to be more conspicuous in *M. tominii*. *Montanelia disjuncta* may also be confused for *M. sorediata*, which occurs elsewhere in Colorado in open granitic habitats. However, the latter species lacks pseudocyphellae and has much finer, powdery soredia that do not become confluent centrally.

CHEMISTRY: Perlatolic acid, stenosporic acid; spot tests: K−, C−, KC−, sometimes KC+ light red (medulla), P−, UV+ white.

SUBSTRATE AND ECOLOGY: Saxicolous on exposed granite and sandstone.

DISTRIBUTION: Relatively common throughout the western United States, Great Lakes region, and New England; also common in montane and boreal Canada and Europe.

LITERATURE: Divakar, P. K., R. Del-Prado, H. T. Lumbsch, M. Wedin, T. L. Esslinger, S. D. Leavitt, and A. Crespo. 2012. "Diversification of the Newly Recognized Lichen-Forming Fungal Lineage *Montanelia* (Parmeliaceae, Ascomycota) and Its Relation to Key Geological and Climatic Events." *American Journal of Botany* 99: 2014–26.

#4847 (PHOTO: J. LENDEMER)

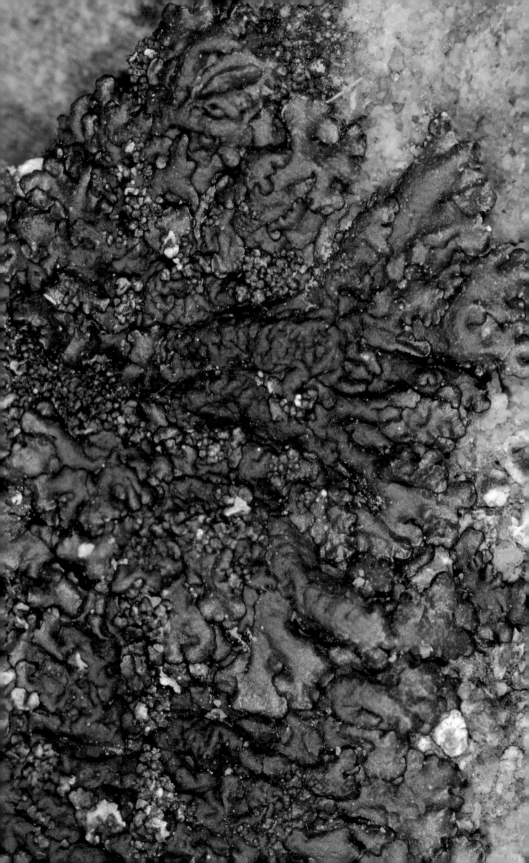

Montanelia tominii

Cacao River

Montanelia tominii is a common saxicolous species of White Rocks and the Front Range of the southern Rocky Mountains. It is one of two species in the genus known from White Rocks, the other being *M. disjuncta* from which it differs by its bright red C+ reaction in the medulla. The pseudocyphellae also tend to be more prominenet in *M. tominii* than in *M. disjunta*. The combination of the C+ reaction and pseudocyphellae makes it impossible to confuse *M. disjuncta* for any other species within the genus or related genera in the "brown Parmelioid" group.

CHEMISTRY: Gyrophoric acid, ovoic acid (minor); spot tests: K–, C+ bright red (medulla), KC+ red (medulla) or KC–, P–, UV–.

SUBSTRATE AND ECOLOGY: On non-calcareous rocks in exposed, arid areas.

DISTRIBUTION: Very common in the western United States and northern Great Plains and the Great Lakes area; also common in montane or boreal Canada and Europe, and reported from northeast Asia.

LITERATURE: Esslinger, T. L. 1977. "A Chemosystematic Revision of the Brown *Parmeliae*." *Journal of the Hattori Botanical Laboratory* 42: 1–211. See also Blanco et al. (2004); Divakar et al. (2012).

#4842 (PHOTO: J. LENDEMER)

Phaeophyscia nigricans

Sandy Oysters

As foliose lichens go, *Phaeophyscia nigricans* is among the more challenging you will encounter. Its thalli, which grow to approximately two cm in diameter, are characteristically grayish-brown to brown. *Phaeophyscia nigricans* is one of just a few species in the genus with pale undersides (a feature otherwise typical of related genera *Anaptychia*, *Physcia* [in part], and *Physciella*). From *Anaptychia* and *Physciella*, *Phaeophyscia nigricans* is differentiable by anatomical features of the upper and lower cortex (paraplectenchymatous in *P. nigricans* versus prosoplectenchymatous in the others). From *Physcia* it is differentiable by its lack of atranorin and thus K– reaction (most *Physcia* are K+ yellow, containing atranorin). In addition to its browinsh color, *P. nigricans* can be identified by its sparse, irregular, isidioid soredia and, perhaps most important, its sparse to numerous one-cell-wide "microhairs" ($\sim$10–12 $\times$ 2–2.5 μM) that occupy the lobe ends. In our study area, *Phaeophyscia nigricans* is perhaps most likely to be confused with *Phaeophyscia hirsuta*, which is similarly pale underneath but has multicellular cortical hairs rather than much smaller microhairs one-cell-width in size. *Phaeophyscia nigricans* was seen only once at White Rocks, occupying the undersides of a rock overhang, and was in a very poorly developed state. This entire specimen appears to have been "sandblasted," thus making an already challenging species to identify even more vexing.

CHEMISTRY: No substances; spot tests: K–, C–, KC–, P–, UV–.

SUBSTRATE AND ECOLOGY: Most commonly on bark, less often on rock.

DISTRIBUTION: Widely distributed in the western United States (rarer in the Pacific Northwest) and upper Great Plains; also known from the Baja Peninsula of Mexico and Europe, particularly northern regions.

LITERATURE: Esslinger, T. L. 2007. "Phaeophyscia." In T. H. Nash III, B. D. Ryan, C. Gries, and F. Bungartz, eds, *Lichen Flora of the Greater Sonoran Desert Region*, vol. 2. Tempe: Lichens Unlimited, Arizona State University, 403–14.

#4865 (PHOTO: E. TRIPP)

Placidium squamulosum

Desert Epoxy

Placidium squamulosum is easily recognized by its brown peltate, squamulose thalli with perithecia that bear hyaline, simple spores. The genus *Peltula* similarly has perithecia embedded in peltate, squamulose, brown thalli and occurs in arid habitats of the West, but it differs by its cyanobacterial photobiont (versus coccoid green algal in *Placidium*). At White Rocks, the only other brown squamulose lichen with perithecia is *Endocarpon pallidulum*, but that species is quite scaly in appearance (see the photo) instead of having larger and more separated squamules typical of *Placidium*. Furthermore, spores of *Endocarpon* are large, brown, and muriform and are embedded in a hymenium that contains algae (which is intriguing and very cool; make a thin section of the perithecia using a razor blade to experience this). Finally, species in the genus *Dermatocarpon* are also peritheciate but form much larger, umbilicate thalli and are generally gray in appearance.

CHEMISTRY: No substances; spot tests: K–, C–, KC–, P–, UV–.

SUBSTRATE AND ECOLOGY: Extremely common on arid soils, both calcium-rich and calcium-poor. Less common on rock.

DISTRIBUTION: *Placidium squamulosum* is one of the most common lichens in western North America. It is arguably the most important lichen species among biotic soil crust communities west of the Mississippi River. In eastern North America, it is the most common member of the genus.

LITERATURE: Breuss, O. 2007b. "Placidium." In T. H. Nash III, B. D. Ryan, C. Gries, and F. Bungartz, eds., *Lichen Flora of the Greater Sonoran Desert Region*, vol. 1. Tempe: Lichens Unlimited, Arizona State University, 384–93.

#4882 (PHOTO: E. TRIPP)

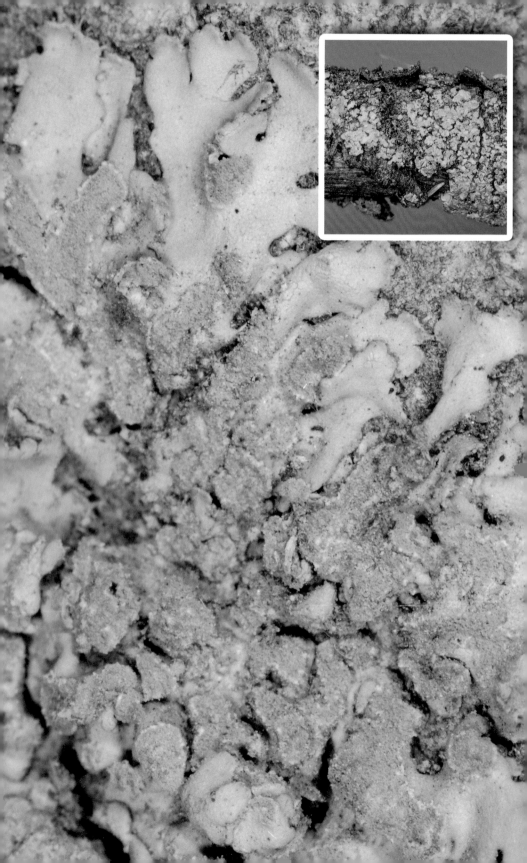

Rib-Ticklers

Physciella melanchra is easily identified by its small, gray, foliose rosettes with abundant soredia and corticolous habit. At White Rocks, it cannot be mistaken for anything else. Elsewhere in our area, *P. melanchra* may be mistaken for the closely related *P. chloantha*, but the latter has terminal, often labriform soredia versus laminal and submarginal irregular soredia as seen in the main photo. Don't waste another minute before obtaining a copy of Esslinger (1986), which provides an excellent summary of features that differentiate *Physciella*, *Physcia*, and *Phaeophyscia*.

CHEMISTRY: No substances; spot tests: K–, C–, KC–, P–, UV–.

SUBSTRATE AND ECOLOGY: Most commonly corticolous (e.g., on *Populus*, *Rhus*, *Celtis*) but occasionally on rock. While *Physciella melanchra* is one of the "weediest" lichens on bark at White Rocks (together with *Xanthomendoza fallax*, see inset), it is actually quite rare across Colorado, and the collection cited above likely represents a new record for Boulder County. The presence of *Physciella melanchra* and numerous other species at White Rocks supports the notion that this sandstone outcropping represents an important island of lichen diversity, much of which is otherwise poorly represented in the state.

DISTRIBUTION: Primarily a species of the Great Plains, extending into suitable habitat in the western and eastern United States; also know from the Sierra Madre Occidental of Mexico and Japan.

LITERATURE: Esslinger, T. L. 1986. "Studies in the Lichen Family Physiaceae. VII. The New Genus *Physciella*." *Mycologia* 78: 92–97.

#4854, 4871 (PHOTOS: E. TRIPP)

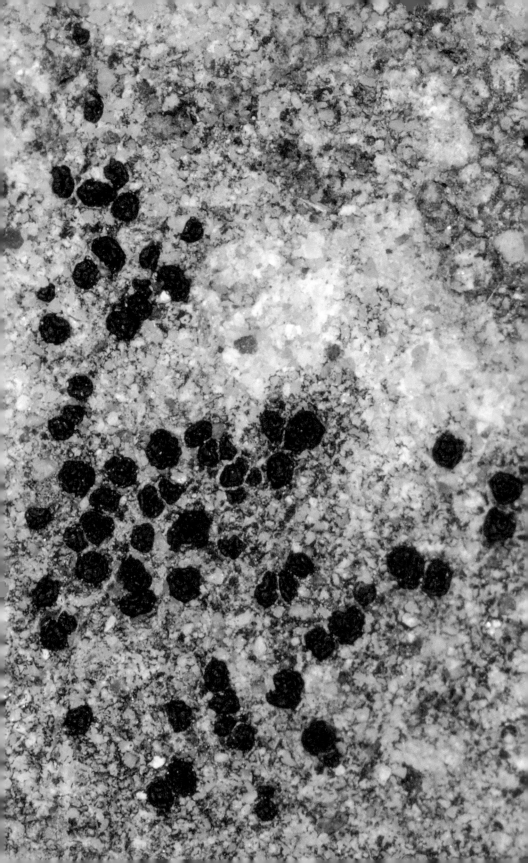

Polysporina simplex

Desperate Dots

A handful of western genera have polysporous asci (i.e., > eight spores/ascus). A lesser number of these have indistinct, endolithic thalli. Of these, *Polysporina* is relatively easy to differentiate from close relatives such as *Sarcogyne* by its black, gyrose discs (as seen in the main photo) that have a carbonized epihymenium (versus brown in *Sarcogyne*). Within *Polysporina*, *P. simplex* is separated from congeners by apothecia width and spore size. This is a very small lichen (discs < 0.6 mm wide), but its black, gyrose discs help identify it tentatively in the field.

CHEMISTRY: No substances; spot tests: K–, C–, KC–, P–, UV–.

SUBSTRATE AND ECOLOGY: Common on acidic rocks, generally in open areas; *Polysporina simplex* is the most common species in the genus in North America.

DISTRIBUTION: Fairly widespread in the United States and particularly abundant in Southern California and the Sonoran region, the Ozarks, central Georgia to Alabama, and the northeastern Great Lakes; also in northern Canada and Europe.

LITERATURE: Knudsen, K. 2007. "Polysporina." In T. H. Nash III, B. D. Ryan, C. Gries, and F. Bungartz, eds., *Lichen Flora of the Greater Sonoran Desert Region*, vol. 3. Tempe: Lichens Unlimited, Arizona State University, 276–78.

#4841 (PHOTO: J. LENDEMER)

Psora tuckermanii

Jasper Squamules

Psora tuckermanii is a common squamulose lichen at White Rocks. It is easily recognized by its dispersed to adjacent brown squamules that are frosty white along the margins as a result of dense pruina (this is variable within the species at other locations). Within the genus, species concepts rely on squamule color, the position of apothecia (on laminar or marginal surfaces), and thallus chemistry. In particular, apothecia position can be challenging to discern; however, only one species of *Psora* is currently known from White Rocks. Elsewhere, the genus *Psora* may be confused for *Placidium* or *Psorula*, but *Placidium* is a peritheciate lichen (versus apotheciate in *Psora*), and *Psorula* has a green lower surface and grows on top of cyanobacterial mats. At White Rocks, *P. tuckermanii* is perhaps most likely to be confused with another brown squamulose lichen, *Endocarpon pallidulum*, but the latter species also produces perithecia rather than apothecia.

CHEMISTRY: No substances; spot tests: K–, C–, KC–, P–, UV–.

SUBSTRATE AND ECOLOGY: *Psora tuckermanii* occurs on both soil and rock substrates. At White Rocks, several thalli were seen covering large areas, indicating a clear ecological importance of this biotic soil crust.

DISTRIBUTION: Very common in western North America, also in southern Canada and northern Mexico.

LITERATURE: Timdal, E. 1986. "A Revision of *Psora* (Lecideaceae) in North America." *Bryologist* 89: 253–75.

#4804, 4879 (PHOTOS: E. TRIPP, V. DÍAZ)

Rhizocarpon disporum

Purple Prose

Rhizocarpon disporum is a crustose, areolate species that is rather distinctive in color among other lichens in our area. The species is easily recognized (and indeed important to learn in Boulder County) because of its brownish-purple thallus, its huge (~80 μM long), brown muriform spores, and its one-spored asci (a most unfortunate specific epithet). The species is also characterized by a K+ purple epihymenium and a conspicuous black prothallus (see top of photo toward margins of thallus). The combination of these features confirms the identification of this species as *R. disporum*. *Rhizocarpon* is a diverse and very important genus in western North America. It is most likely to be confused with *Buellia*, *Porpidia*, or *Lecidea*, but species of *Buellia* lack halos around their spores (versus halonate in *Rhizocarpon*), species of *Porpidia* have one-celled spores inside a different ascus type (versus two- to many-celled spores in *Rhizocarpon*), and species of *Lecidea* have small, one-celled hyaline spores and unbranched paraphyses (versus brown spores and branched paraphyses in *Rhizocarpon*).

CHEMISTRY: With or without norstictic acid; spot tests: K– or K+ red (cortex when norstictic acid is present), K+ purple (epihymenium), C–, KC–, P– or P+ red (cortex), UV–.

SUBSTRATE AND ECOLOGY: Exclusively saxicolous, especially on sandstone and granite. This taxon is uncommon at White Rocks but very common elsewhere in the foothills of Boulder County.

DISTRIBUTION: Widespread in western North America, the northern Great Plains, and the Great Lakes region; also known from numerous other areas worldwide.

LITERATURE: Feuerer, T., and E. Timdal. 2007. "Rhizocarpon." In T. H. Nash III, B. D. Ryan, C. Gries, and F. Bungartz, eds., *Lichen Flora of the Greater Sonoran Desert Region*, vol. 2. Tempe: Lichens Unlimited, Arizona State University, 456–66.

#4861 (PHOTO: E. TRIPP)

Rhizoplaca chrysoleuca

Rocky Mountain High

Rhizoplaca chrysoleuca is an easy-to-identify species and in fact is one of the most common saxicolous lichens in Boulder County and much of the southern Rockies. It is recognizable by its umbilicate thallus and hyper-fertile state in which apothecia discs are salmon-orange except when extremely immature (see upper-right portion of photo). The thallus is distinctively greenish-yellow because of the presence of usnic acid. These features (umbilicate, salmon-orange discs, greenish-yellow thallus) combined with placodiolic or pseudoplacodiolic acid in the cortex as determined by thin layer chromatography readily identify *R. chrysoleuca*. In fact, the genus as a whole is generally difficult to mistake for anything else. Elsewhere in our area, this species is most likely to be confused with the *Rhizoplaca melanompthalma* complex, but the latter have dark apothecia discs. *Rhizoplaca* is closely related to *Lecanora* in the Lecanoraceae (Miadlikowska et al. 2014), with which it shares lecanorine apothecia, *Lecanora*-type asci, and small, simple, hyaline spores.

CHEMISTRY: Usnic acid (cortex), placodiolic or pseudoplacodiolic acid (cortex), aliphatic acids (sometimes in medulla), psoromic acid (occasional), lecanoric acid (occasional); spot tests: K+ yellow (cortex), C–, KC+ yellow (cortex) or KC+ red (medulla), P– or P+ yellow (medulla), UV–.

SUBSTRATE AND ECOLOGY: Very frequent on siliceous rocks, infrequent on calcareous rocks; this species is very wide-ranging ecologically, from dry, arid piñon-juniper scrublands to alpine environments.

DISTRIBUTION: Extremely common in the western United States and upper Great Plains; also known in Canada, Europe, Asia, and South America.

LITERATURE: Ryan, B. D. 2007. "Rhizoplaca." In T. H. Nash III, B. D. Ryan, C. Gries, and F. Bungartz, eds., *Lichen Flora of the Greater Sonoran Desert Region*, vol. 1. Tempe: Lichens Unlimited, Arizona State University, 442–48.

#4816 (PHOTO: E. TRIPP)

Rinodina pyrina

Bark Pots

Rinodina is an easy taxon to identify to genus but a somewhat challenging group to identify to species, which requires extensive study of spore anatomy and development. The genus is characterized by its crustose, continuous to areolate thalli that are often green to brown and very small in size, its darkened apothecia with lecanorine margins, and its brown, two-celled (rarely four-celled) spores. The genus is perhaps most commonly confused with *Buellia*, from which it is easily differentiable by the presence of a proper thalline margin (composed of hyphal tissue derived from the thallus), its unpigmented hypothecium, and a specific ascus type. *Rinodina pyrina* is one of three known species in the genus at White Rocks. Fortunately, it is easily separable from the other two by its corticolous habit. *Rinodina pyrina* is characterized by its gray thallus, crowded apothecia, and *Physconia*-type spores (see Sheard 2010). In figure at left, *R. pyrina* is pictured among much larger thalli of *Physciella melanchra*. Each fruiting body of *R. pyrina* is < 0.5 mm in diameter. *Rinodina* is extremely diverse in North America. For a treatment of North American species on the whole, see Sheard's excellent monograph cited below. For an updated treatment of the southeastern species, see Lendemer et al. 2014.

CHEMISTRY: No substances; spot tests: K–, C–, KC–, P–, UV–.

SUBSTRATE AND ECOLOGY: Corticolous on a diversity of shrubs and trees.

DISTRIBUTION: Fairly widespread in western North America; also reported from Europe and Asia.

LITERATURE: Sheard, J. W. 2010. *The Lichen Genus* Rinodina *(Ach.) Gray (Lecanoromycetidae, Physciaceae) in North America, North of Mexico.* Ottawa: NRC Research Press.

#4872 (PHOTO: E. TRIPP)

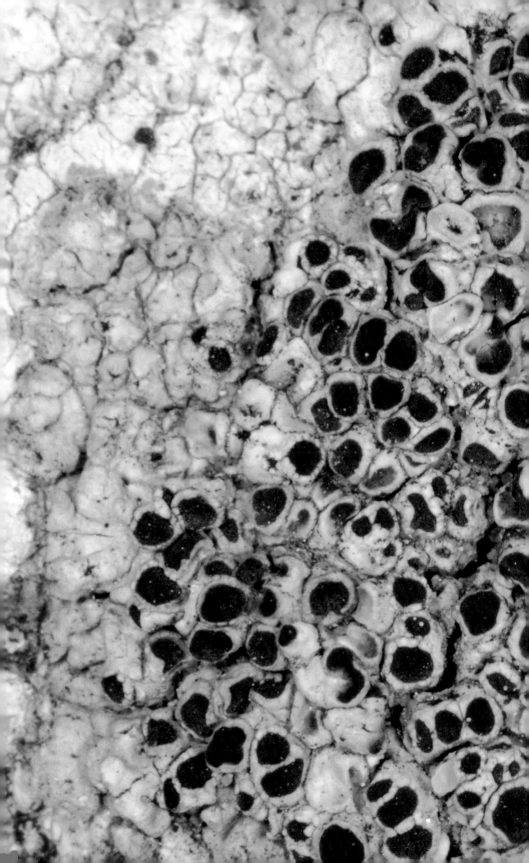

Rinodina straussii

Ink Pots

Rinodina straussii is one of two saxicolous species in the genus at White Rocks (the other being *R. venostana*). It is identifiable by its thick, sordid white to light gray areoles that are sometimes stalked in appearance and its large black discs that are often pruinose. In contrast, *R. venostana* has a thinner, grayer thallus. It is imperative to learn the intricacies of spore morphology if one is to identify species of *Rinodina* with any confidence. See Sheard (2010) for much more extensive information on this subject and on *R. straussii*.

CHEMISTRY: No substances; spot tests: K–, C–, KC–, P–, UV–.

SUBSTRATE AND ECOLOGY: On limestone as well as sandstone.

DISTRIBUTION: Endemic to the western United States, from Colorado to southern California.

LITERATURE: Sheard, J. W. 2010. *The Lichen Genus* Rinodina *(Ach.) Gray (Lecanoromycetidae, Physciaceae) in North America, North of Mexico.* Ottawa: NRC Research Press.

#4814 (PHOTO: E. TRIPP)

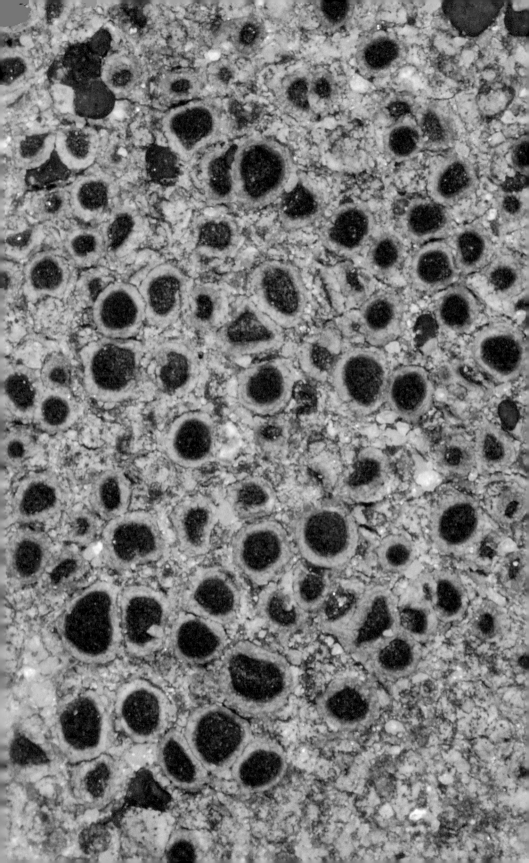

Rinodina venostana

Sandstone Pucks

Rinodina venostana is one of two saxicolous species in the genus at White Rocks. It is generally distinguished from the other, *R. straussii*, by its thin, gray thallus (versus thicker, sordid white areolae in the latter). However, thallus color and degree of development can be variable in these and other species of *Rinodina*; thus, learning technical details of sectioning and studying spore type are essential for identifying species of *Rinodina*.

CHEMISTRY: No substances; spot tests: K–, C–, KC–, P–, UV–.

SUBSTRATE AND ECOLOGY: This species occurs exclusively on very soft and fragile substrates such as sandstones and schists. Keep your eyes open for it while visiting similar ecosystems throughout the western United States.

DISTRIBUTION: This collection represents the first report of this species in the United States. In fact, it was not known from North American until the 2014 report from Saskatchewan, cited below. The type collection was made in Italy, and the species has also been reported from southwestern Germany.

LITERATURE: Freebury, C. E. 2014. "Lichens and Lichenicolous Fungi of Grassland National Park (Saskatchewan, Canada)." *Opuscula Philolichenum* 13: 102–21.

#4858 (PHOTO: E. TRIPP)

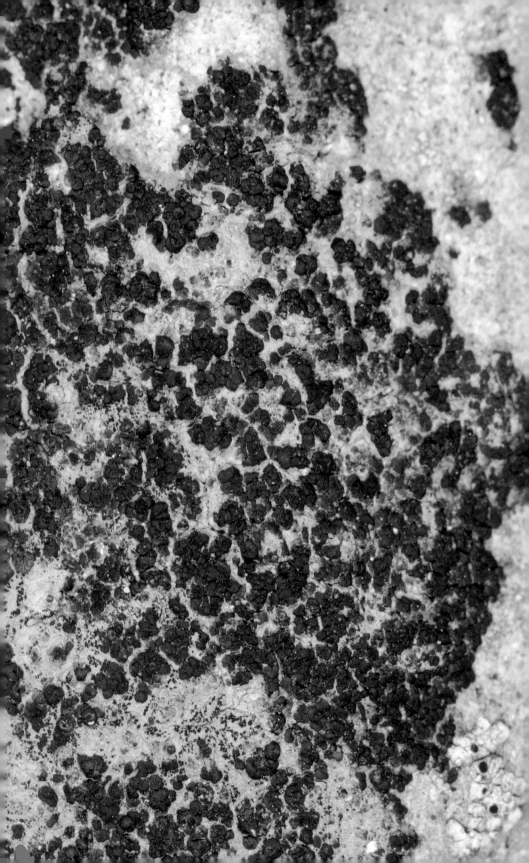

Staurothele areolata

Chocolate Spackle

Staurothele areolata is an extremely common crustose lichen at White Rocks. It is recognizable by its very small, dark brown areoles, the larger ones of which contain one large, brownish-black perithecium toward the areole center, and its brown muriform spores inside a hymenium that contains algae. At White Rocks, *S. areolata* can only be confused with *Endocarpon pallidulum*, which also has brown, muriform spores and algae present within the hymenium. However, *S. areolata* differs from *E. pallidulum* by its squamules that arise from the cracking of a continuous thallus (versus arising individually from a prothallus, as in *Endocarpon*). A quick check for a lower cortex and rhizine-like attachment structures in *Endocarpon* will help confirm that genus. In addition, thalli of *S. areolata* tend to be shiny and darker brown versus the duller, lighter brown thalli of *Endocarpon*. Gestalt works well too: *S. areolata* and *E. pallidulum* honestly don't look anything alike. *Heteroplacidium zamenhofianum* is a crustose species that commonly parasitizes *S. areolata* but hasn't yet been found at White Rocks.

CHEMISTRY: No substances; spot tests: K–, C–, KC–, P–, UV–.

SUBSTRATE AND ECOLOGY: Very common on silicaceous as well as basic rocks.

DISTRIBUTION: Widespread in western North America and the north-central Great Plains; also reported elsewhere worldwide.

LITERATURE: Thomson, J. W. 2007. "Staurothele." In T. H. Nash III, B. D. Ryan, C. Gries, and F. Bungartz, eds., *Lichen Flora of the Greater Sonoran Desert Region*, vol. 1. Tempe: Lichens Unlimited, Arizona State University, 468–72.

#4802 (PHOTO: E. TRIPP)

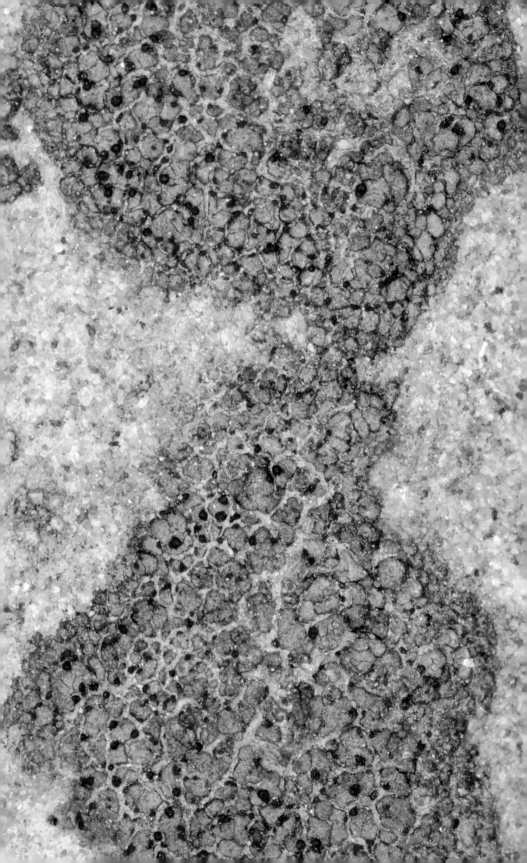

Verrucaria beltraminiana

Outlaw Ashes

Verrucariaceae is not exactly your friendly neighborhood grocer. But you're the one who decided to live in the West, and just like those terrible Chinook winds, now you get to deal with Verrucariaceae (secret: I am quite fond of both). With ~61 species, *Verrucaria* is well represented in our region and is characterized by its always crustose habit, its perithecia that are immersed in the thallus or substrate (these sometimes superficial) and arranged either laminally or distinctly between the areoles, its eight-spored asci with simple hyaline spores, and an almost always saxicolous substrate. To identify species, one must section mature perithecia with a straight cut through the ostiole; cutting on an angle will distort important characters relating to the involucrellum, such as thickness. See lovely drawings of examples of involucrellum diversity in *Verrucaria* by Othmar Breuss on pages 360–61 of volume 3 of *Lichen Flora*, cited below. Spore size is also very important in identifying *Verrucaria*, so don't try it without a compound microscope. *Verrucaria beltraminiana* is characterized by its typical gray thallus with perithecia generally situated toward the edges of the areoles (see the photo), but Breuss (2007c) notes that spore size and perithecium dimensions are the only reliable means of identifying the species, especially against the rich diversity of congerns in the southwestern United States. At White Rocks, it is most likely to be confused with *V. glaucovirens*, but the latter has subdivided areolae, unlike the continuous areolae of *V. beltraminiana*.

CHEMISTRY: No substances; spot tests: K–, C–, KC–, P–, UV–.

SUBSTRATE AND ECOLOGY: Saxicolous on both calcareous and non-calcareous rocks.

DISTRIBUTION: Previously known from Europe (type locality), Arizona, and Mexico. The above collection represents the first confirmed record for Colorado.

LITERATURE: Breuss, O. 2007c. "Verrucaria." In T. H. Nash III, B. D. Ryan, C. Gries, and F. Bungartz, eds., *Lichen Flora of the Greater Sonoran Desert Region*, vol. 3. Tempe: Lichens Unlimited, Arizona State University, 335–77.

#4810 (PHOTO: E. TRIPP)

Verrucaria furfuracea

On the Rocks

Unlike *Verrucaria beltraminiana* and *V. glaucovirens*, whose identities may vex you, *V. furfuracea* is very easily identified by brown, epruinose areolae that boast highly conspicuous isidia along the margins. *Verrucaria furfuracea* also has a thick, subsquamulose thallus and an incomplete involucrellum. To my knowledge, this is the only isidiate *Verrucaria* in our immediate area or, more broadly, across western North America. See an additional photo of this species in Breuss's (2007c) treatment of the genus in *Lichen Flora*. When all else fails in your quest to learn crustose (and, at first, inconspicuous) lichens, try the likability test. The excitement of *Verrucaria* isidia once made me spill an entire martini (with Hendrick's Gin, no less) on my home microscope.

CHEMISTRY: No substances; spot tests: K–, C–, KC–, P–, UV–. Best to move on to other genera if you're interested in chemical diversity.

SUBSTRATE AND ECOLOGY: Saxicolous on both calcareous and non-calcareous rocks.

DISTRIBUTION: Known from semiarid areas in North and South Dakota through Kansas, Colorado, Utah, Arizona, and Southern California. In Colorado, it has been reported from El Paso and Kit Carson Counties.

LITERATURE: Breuss, O. 2007c. "Verrucaria." In T. H. Nash III, B. D. Ryan, C. Gries, and F. Bungartz, eds., *Lichen Flora of the Greater Sonoran Desert Region*, vol. 3. Tempe: Lichens Unlimited, Arizona State University, 335–77.

#4831 (PHOTO: E. TRIPP)

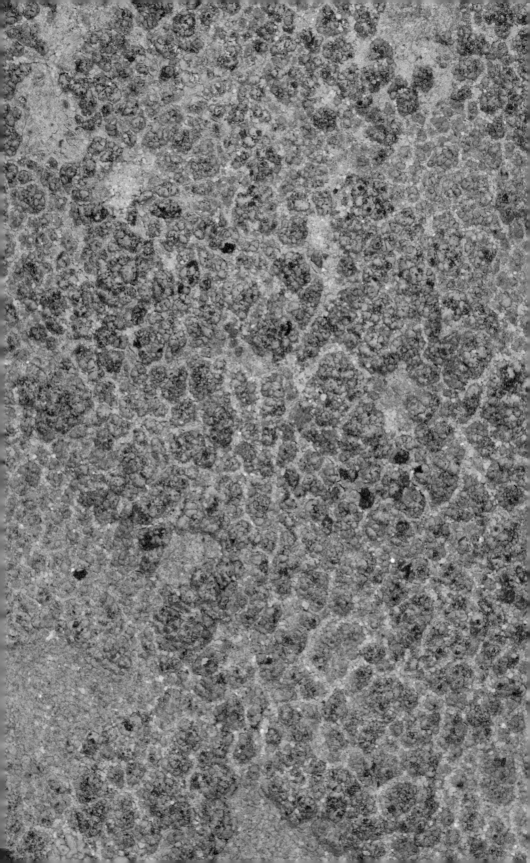

Verrucaria glaucovirens

Lunar Walk

Verrucaria glaucovirens is characterized by its thick and basally constricted subdivided areolae, which is discernible in figure at left. The subdivided areole of *V. glaucovirens* represent the easiest way to differentiate the species from *V. beltraminiana*, which it most closely resembles at White Rocks and which has areoles that are usually contiguous. See information under *V. beltraminiana* for a general overview of the genus in North America.

CHEMISTRY: No substances; spot tests: K–, C–, KC–, P–, UV–.

SUBSTRATE AND ECOLOGY: Always saxicolous on sandstone.

DISTRIBUTION: The distribution of this species is incompletely known, but it seems to track semiarid sandstone formations across the southern Rocky Mountains as well as the upper Midwest and Great Lakes.

LITERATURE: Breuss, O. 2007c. "Verrucaria." In T. H. Nash III, B. D. Ryan, C. Gries, and F. Bungartz, eds., *Lichen Flora of the Greater Sonoran Desert Region*, vol. 3. Tempe: Lichens Unlimited, Arizona State University, 335–77.

#4851 (PHOTO: E. TRIPP)

Xanthomendoza fallax

A Typical Monday

Xanthomendoza fallax is one of three very common (dare we say "weedy") corticolous macrolichens at White Rocks (the other two are *Physciella melanchra* and *X. galericulata*). *Xanthomendoza fallax* is easily identified by its small, rosette-forming thalli that are loosely connected to the substrate by abundant rhizines and the presence of soredia that form in crescent shaped slits (soralia) between the upper and lower cortex. At White Rocks, *X. fallax* is differentiated from *X. galericulata* by its distinct greenish-yellow thallus (versus thalli that are typically bright orange in *X. galericulata*; elsewhere, the two species can appear more similar in color but can be differentiated by other features, as described in the key). *Xanthomendoza fallax* is one of several sorediate species in the genus that are present in western North America, which can be differentiated based on features of the thallus, rhizines, and soralia (see citation below as well as Lindblom 2006, 2007).

CHEMISTRY: Emodin, fallacinal, parietin, parietinic acid, teloschistin; spot tests: K+ purple (cortex), C–, KC–, P–, UV–.

SUBSTRATE AND ECOLOGY: Common on corticolous substrates; rare on rock or other surfaces.

DISTRIBUTION: Very widespread in western North America, the Great Plains, and portions of eastern North America (especially the Great Lakes region); also present in several other areas worldwide.

LITERATURE: Søchting, U., I. Kärnefelt, and S. Kondratyuk. 2002. "Revision of *Xanthomendoza* (Teloschistaceae, Lecanorales) Based on Morphology, Anatomy, Secondary Metabolites and Molecular Data." *Mitteilungen aus dem Institut für Allgemeine Botanik in Hamburg* 30–32: 225–40.

#4853 (PHOTO: E. TRIPP)

Xanthomendoza galericulata

Frilly Drawers

Xanthomendoza galericulata is one of two corticolous species in the genus at White Rocks and is distinguishable from the other species, *X. fallax*, by its distinctly dark orange thallus (greenish-yellow in *X. fallax*), relatively sparse rhizines (denser in *X. fallax*), and granular soredia that form in helmet-shaped soralia at lobe tips that appear almost inflated (marginal soralia in slits in *X. fallax*). For distinctions between *X. galericulata* and other close relatives, see a key to the sorediate species of *Xanthomendoza* in North America in Lindblom (2006). This is a lovely little lichen that you can learn on day one at White Rocks or anywhere else.

CHEMISTRY: Parietin and teloschistin (major), emodin, fallacinal, parietinic acid (minor); spot tests: K+ purple (cortex and apothecia), C–, KC–, P–, UV–.

SUBSTRATE AND ECOLOGY: Occurring on a wide variety of corticolous substrates of western North America; relatively common in arid areas.

DISTRIBUTION: Fairly widespread from western Colorado north and west to eastern Washington and south to Southern California.

LITERATURE: Lindblom, L. 2006. "*Xanthomendoza galericulata*, a New Sorediate Lichen Species, with Notes on Similar Species in North America." *Bryologist* 109: 1–8.

#4827 (PHOTO: J. LENDEMER)

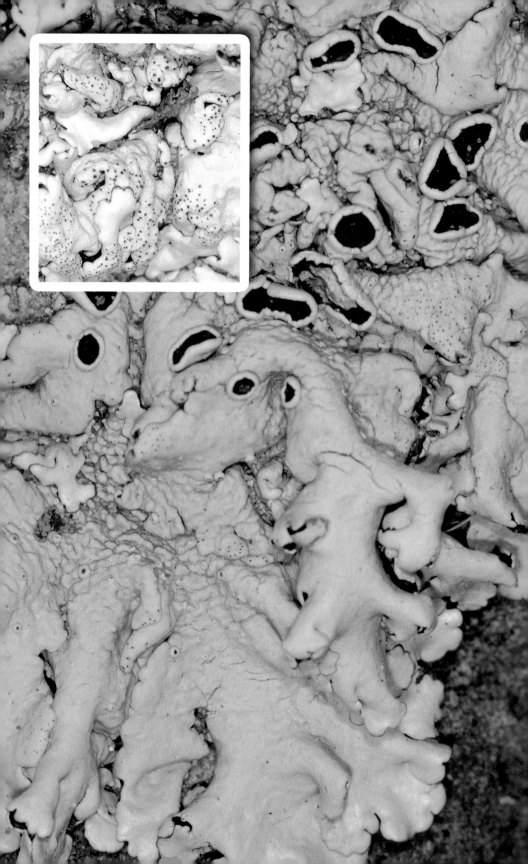

Xanthoparmelia coloradoensis

Geyser Sludge

Within *Xanthoparmelia*, species display one of two primary reproductive strategies: sexual (apothecia) or asexual (isidia or soredia), and these types exist in more or less equal frequencies throughout the West. Although species of *Xanthoparmelia* are large macrolichens, identification can be challenging, and thin layer chromatography is a must. Lower cortex color and degree of thallus attachment to the substrate are important characters during species identification. *Xanthoparmelia coloradoensis* is distinguished by its pale lower surface, chemistry, loose adnation to rock substrates, and broad lobes. In our area, this taxon is most likely to be confused with *X. lineola*, which has a similar chemistry and lower cortex color but differs by being more closely adnate to rocks and thus impossible to "peel" from the substrate. Be aware that *Xanthoparmelia coloradoensis* is commonly encountered only with pycnidia (inset photo), lacking apothecia seen in the main photo. *Xanthoparmelia* is one of the most diverse and ecologically important genera in western North America, especially in Colorado, where the genus needs revision.

CHEMISTRY: Consalazinic acid, salazinic acid, usnic acid (major), and norstictic acid, protocetraric acid (trace); spot tests: K+ yellow to red (medulla), KC–, C–, P+ orange (medulla), UV–.

SUBSTRATE AND ECOLOGY: Very common on soil and non-calcareous rocks in open, arid environments.

DISTRIBUTION: Widespread in western North America, including Canada; also in other areas worldwide.

LITERATURE: Nash, T. H., III, and J. A. Elix. 2007. "Xanthoparmelia." In T. H. Nash III, B. D. Ryan, C. Gries, and F. Bungartz, eds., *Lichen Flora of the Greater Sonoran Desert Region*, vol. 2. Tempe: Lichens Unlimited, Arizona State University, 566–605.

#4840, 4855 (PHOTOS: E. TRIPP, J. LENDEMER)

Xanthoparmelia lavicola

The Trochanter Lichen

Xanthoparmelia lavicola is one of several isidiate species in the genus in our area. It is distinguished from others by having isidia that are conspicuously globose (see inset photo) versus strongly cylindrical (as in *X. plittii*, not yet reported from White Rocks), by its chemistry (versus that of *X. mexicana*), and by its pale lower cortex (black in *X. conspersa*). In addition, lobes of *Xanthoparmelia lavicola* tend to be rather large in size. Thin layer chromatography is essential to identify species of *Xanthoparmelia* with accuracy. *Xanthoparmelia lavicola* is one of only two species in the genus in Colorado that produces psoromic acid; the other, *X. psoromifera*, is only known from the southeastern portion of the state at present. *Xanthoparmelia* is very diverse and important ecologically in western North America, albeit still poorly understood in several regions such as Colorado, where we as yet do not fully understand how species that are very similar morphologically occupy different niches ecologically, as of course they do.

CHEMISTRY: Psoromic acid, usnic acid (major), 2'-*O*-demethylpsoromic acid (minor), subpsoromic acid (trace); spot tests: K–, C–, KC–, P+ yellow (medulla), UV–.

SUBSTRATE AND ECOLOGY: On soil and non-calcareous rocks in open, arid environments such as White Rocks.

DISTRIBUTION: Southern Rocky Mountains and desert southwestern United States, extending into Mexico along the Sierra Madre Occidental and Oriental; its center of distribution of southwestern Arizona and northeastern Mexico.

LITERATURE: Nash, T. H., III, and J. A. Elix. 2007. "Xanthoparmelia." In T. H. Nash III, B. D. Ryan, C. Gries, and F. Bungartz, eds., *Lichen Flora of the Greater Sonoran Desert Region*, vol. 2. Tempe: Lichens Unlimited, Arizona State University, 566–605.

#4839 (PHOTO: J. LENDEMER)

Xanthoria elegans (Link) Th. Fr.

Orange Spectacle

Xanthoria elegans is one of the most common saxicolous lichens in Boulder County but is comparatively less abundant at White Rocks. This difference is likely attributable to its affinity for nutrient-rich substrates, such as calcareous rock or mortar or rock enriched by animal urine (versus nutrient-poor surfaces that characterize this sandstone formation). At White Rocks, *Xanthoria elegans* might be confused with several species of *Caloplaca* that have small to moderately sized thalli, such as *C. saxicola* or *C. subsoluta*, but it differs by having a lower cortex and is thus removable intact from rock substrates. The thallus of *Caloplaca trachyphylla* is similar in general appearance but has much longer lobes (1.5–5 mm versus usually < 2 mm long in *X. elegans*) and is much larger in size than *X. elegans*. Elsewhere, *X. elegans* might be confused with *X. parietina*, but the latter species is primarily corticolous in Colorado and differs in other features pertaining to the lobes. Make this species one of the first that you learn—out West, it is EVERYWHERE.

CHEMISTRY: Parietin (major), emodin, fallacinal, parietinic acid, teloschistin (minor); spot tests: K+ purple (cortex), C–, KC–, P–, UV–.

SUBSTRATE AND ECOLOGY: On a variety of rock surfaces, particularly mineral-rich surfaces and those fertilized by animals (e.g., near dens).

DISTRIBUTION: Extremely common throughout the western and northeastern United States and Canada; also present on most other continents.

LITERATURE: Lindblom, L. 2007. "Xanthoria." In T. H. Nash III, B. D. Ryan, C. Gries, and F. Bungartz, eds., *Lichen Flora of the Greater Sonoran Desert Region*, vol. 2. Tempe: Lichens Unlimited, Arizona State University, 605–11.

#4843 (PHOTO: J. LENDEMER)

Appendix

Checklist of the Lichens of White Rocks

A complete set of all voucher specimens cited below is housed in the University of Colorado (COLO) Herbarium. In Tripp (2015), taxon abundance was assessed and measured on a scale with five rankings—rare, infrequent, occasional, common, abundant—and these rankings are shown below. Notes specific to a taxon report are also provided.

Taxon	Collection Number(s)	Abundance Scale	Notes
Acarospora obpallens (Nyl. ex Hasse) Zahlbr.	E. Tripp and D. Clark 4817; E. Tripp et al. 4830	occasional	
Acarospora sp. nov.	E. Tripp 4820	rare	1
Acarospora stapfiana (Müll. Arg.) Hue	E. Tripp 4806	occasional	
Acarospora strigata (Nyl.) Jatta	E. Tripp et al. 4559; E. Tripp 4808, 4868	abundant	
Aspicilia cinerea (L.) Körber	E. Tripp and V. Díaz 4880	occasional	
Caloplaca atroflava (Turner) Mong.	E. Tripp and D. Clark 4824	abundant	2
Caloplaca decipiens (Arnold) Blomb. and Forss.	E. Tripp et al. 4845	occasional	
Caloplaca pratensis Wetmore	E. Tripp et al. 4836; E. Tripp 4873	occasional	
Caloplaca saxicola (Hoffm.) Nordin	E. Tripp et al. 4838	common	
Caloplaca sideritis (Tuck.) Zahlbr.	E. Tripp et al. 4835	infrequent	
Caloplaca subsoluta (Nyl.) Zahlbr.	E. Tripp et al. 4832	occasional	
Caloplaca trachyphylla (Tuck.) Zahlbr.	E. Tripp 4812	abundant	
Candelariella clarkiae E. Tripp and Lendemer	E. Tripp and V. Díaz 4876	rare	3

continued on next page

Taxon	Collection Number(s)	Abundance Scale	Notes
Candelariella rosulans (Müll. Arg.) Zahlbr.	E. Tripp 4809, 4811; E. Tripp and V. Díaz 4877	abundant	
Dermatocarpon americanum Vain.	E. Tripp et al. 4844	infrequent	
Dermatocarpon moulinsii (Mont.) Zahlbr.	E. Tripp et al. 4828	rare	
Diploschistes scruposus (Schreber) Norman	E. Tripp and D. Clark 4857, 4859; E. Tripp 4867	occasional	
Diplotomma venusta (Körb.) Körb.	E. Tripp et al. 4849	occasional	
Enchylium tenax (Sw.) Gray	E. Tripp and V. Díaz 4878	infrequent	
Endocarpon pallidulum (Nyl.) Nyl.	E. Tripp et al. 4829; E. Tripp and D. Clark 4850, 4860, E. Tripp 4862	occasional	
Lecanora argopholis (Ach.) Ach.	E. Tripp and D. Clark 4819, 4823	common	
Lecanora flowersiana H. Magn.	E. Tripp 4815	rare	4
Lecanora garovaglii (Körb.) Zahlbr.	E. Tripp 4813; E. Tripp et al. 4833; E. Tripp and D. Clark 4852	common	
Lecanora muralis (Schreber) Rabenh.	E. Tripp and D. Clark 4822	common	
Lecanora sp. nov.	E. Tripp and D. Clark 4823	rare	5
Lecidea hoganii E. Tripp and Lendemer	E. Tripp 4801, 4805	infrequent	3
Lecidea tessellata Flörke	E. Tripp and D. Clark 4818, 4856	infrequent	6
Lecidella carpathica Körb.	E. Tripp and V. Díaz 4876	infrequent	

continued on next page

Taxon	Collection Number(s)	Abundance Scale	Notes
Lecidella patavina (A. Massal.) Knoph and Leuckert	E. Tripp 4869, 4870	abundant	
Lecidella stigmatea (Ach.) Hertel and Leuckert	E. Tripp et al. 4846	infrequent	
Lecidella viridans (Flot.) Körb.	E. Tripp et al. 4837	infrequent	
Lepraria finkii (B. de Lesd.) R. C. Harris	E. Tripp and D. Clark 4825	common	
Lichinella nigritella (Lettau) Henssen	E. Tripp 4807	infrequent	
Lichinella stipatula Nyl.	E. Tripp 4803b	occasional	
Lobothallia alphoplaca (Wahlenb.) Hafellner	E. Tripp et al. 4834	occasional	
Montanelia disjuncta (Erichsen) Divakar	E. Tripp et al. 4847	rare	
Montanelia tominii (Oxner) Divakar	E. Tripp et al. 4842	occasional	
Phaeophyscia nigricans (Flörke) Moberg	E. Tripp 4865	rare	
Physciella melanchra (Hue) Essl.	E. Tripp and D. Clark 4854; E. Tripp 4871	abundant	
Placidium squamulosum (Ach.) Breuss.	E. Tripp and V. Díaz 4882	occasional	
Polysporina simplex (Taylor) Vězda	E. Tripp et al. 4841	common	
Psora tuckermanii R. A. Anderson ex Timdal	E. Tripp 4804; E. Tripp and V. Díaz 4879	occasional	
Rhizocarpon disporum (Nägeli ex Hepp) Müll. Arg.	E. Tripp and D. Clark 4861	occasional	
Rhizoplaca chrysoleuca (Sm.) Zopf	E. Tripp and D. Clark 4816	common	

continued on next page

Taxon	Collection Number(s)	Abundance Scale	Notes
Rinodina pyrina (Ach.) Arnold	E. Tripp 4872	common	
Rinodina straussii J. Steiner	E. Tripp 4814	occasional	
Rinodina venostana Buschardt and H. Mayrhofer	E. Tripp and D. Clark 4858	rare	7
Staurothele areolata (Ach.) Lettau	E. Tripp 4802, 4803, 4863	abundant	
Verrucaria beltraminiana (A. Massal.) Trevis.	E. Tripp 4810	common	8
Verrucaria furfuracea (B. de Lesd.) Breuss	E. Tripp et al. 4831	infrequent	
Verrucaria glaucovirens Grummann	E. Tripp 4851	infrequent	
Xanthomendoza fallax (Hepp ex Arnold) Søchting, Kärnefelt, and S. Y. Kondr.	E. Tripp and D. Clark 4853	abundant	
Xanthomendoza galericulata L. Lindblom	E. Tripp et al. 4827	abundant	
Xanthoparmelia coloradoensis (Gyelnik) Hale	E. Tripp et al. 4840; E. Tripp and D. Clark 4855	occasional	
Xanthoparmelia lavicola (Gyelnik) Hale	E. Tripp et al. 4839; E. Tripp and V. Díaz 4874	occasional	
Xanthoria elegans (Link) Th. Fr.	E. Tripp et al. 4843	occasional	

Notes

1. An undescribed taxon that is still under investigation.

2. Most material resembling this taxon (i.e., red apothecia, inconspicuous thalli) within North American herbaria is identified as *Caloplaca arenaria* (Pers.) Müll. Arg. The collection from White Rocks has a definitive, albeit thin, gray thallus that reacts K+ purple in water mount and does not appear to be obviously parasitic on other lichens. As such, the name *Caloplaca atroflava* is here applied to this collection instead of *C. arenaria* or *C. epithallina*.

3. The manuscript reporting these two new species includes molecular exploration of evolutionary relationships to close relatives (Tripp and Lendemer 2015).

4. This species is member to the Lecanora dispersa group that is characterized by absent to endolithic (and not visible) thalli. At White Rocks, there exists a second member of this group which is likely undescribed to science. The former can, however, be easily distinguished by its lack of usnic acid (present in the latter).

5. This is an usnic acid–containing species that is not readily ascribable to anything in Śliwa's monograph (Śliwa 2007) of the *Lecanora dispersa* complex.

6. The name *Lecidea tessellata* is unsatisfactory for this taxon; for example, collections from White Rocks have an I- medulla and a dark gray thallus, and they lack a black prothallus. However, until the genus or this species is fully revised in western North America and the many entities contained within *L. tessellata* are separated out, this is the most suitable name for White Rocks material.

7. *Rinodina venostana* was recently reported as new to North America from a sandstone formation in Saskatchewan (Freebury 2014). This collection reported here serves as the first documentation of *R. venostana* in the United States, although there is some possibility that eventually it might be separated out as a taxon new to science (J. Sheard, pers. comm.).

8. This record represents the first confirmed report of this species for Colorado.

Dichotomous Key

to the Lichens of White Rocks

1. Lichens growing on bark 2
1. Lichens growing on rock 5

2. Thallus bright orange or yellow 3
2. Thallus gray to green 4

3. Thallus greenish-yellow; soralia in slits between the upper and lower cortices; soredia mostly greenish-yellow . . . . **Xanthomendoza fallax (p. 125)**
3. Thallus deep orange; soralia helmet- or hood-shaped; soredia mostly orange . . . . **Xanthomendoza galericulata (p. 140)**

4. Crustose; reproducing primarily sexually through ascospores . . . . **Rinodina pyrina (p. 111)**
4. Foliose; reproducing primarily asexually through soredia . . . . **Physciella melanchra (p. 101)**

5. Lichens with a cyanobacterial photobiont (blackish in color) 6
5. Lichens with a green algal photobiont (never black in color) 8

6. Thallus dwarf fruticose . . . . **Lichinella stipatula (p. 89)**
6. Thallus foliose 7

7. Lobes narrow (usually < 2 mm wide at apex), not easily discernible with a hand lens, apex shape not obvious without microscopy; photobiont primarily filamentous . . . . **Enchylium tenax (p. 59)**
7. Lobes broad (usually > 2 mm wide at apex), easily discernible with a hand lens, rounded at apex, and ascending from substrate; photobiont primarily chroococcoid . . . . **Lichinella nigritella (p. 87)**

8. Lichens reproducing primarily sexually through apothecia 9
8. Lichens reproducing primarily sexually through perithecia or asexually . . . 42

9. Thallus and/or fruiting bodies bright orange, yellow, or red 10
9. Thallus and/or fruiting bodies not orange, yellow, or red 18

10. Thallus and/or fruiting bodies shades of yellow, K−; spores simple 11
10. Thallus and/or fruiting bodies orange to red, K+ purple; spores polarilocular 12

11. Thallus weakly formed, of dispersed areoles, mostly visible as small patches of tissue that surround apothecia; chartreuse........ Candelariella clarkiae (p. 47)
11. Thallus of well-developed areoles or squamules extending well beyond the apothecia, neon yellow or yolk yellow but not chartreuse............... 13

12. Thallus neon yellow, often seen growing on thalli of *Caloplaca trachyphylla* and apparently parasitic; asci polysporous (> 100 spores/ascus) ....
.... Acarospora stapfiana (p. 25)
12. Thallus yolk yellow, free-living and not obviously associated with any particular species of lichen; asci with 8–32 spores each . Candelariella rosulans (p. 49)

13. Thallus dark gray, apothecia orange................. Caloplaca sideritis (p. 41)
13. All parts of thallus and/or apothecia orange........................... 14

14. Thallus minute, almost indiscernible.................................. 15
14. Thallus with well-developed squamules or lobes 16

15. Apothecia distinctly orangish-red, thallus minute and very thin, gray ....
.... Caloplaca atroflava (p. 33)
15. Apothecia orange, thallus minute, concolorous with thallus ....
.... Caloplaca subsoluta (p. 43)

16. Thallus with small squamules, these 1–2 mm long... Caloplaca saxicola (p. 39)
16. Thallus with definitive lobes, these > 2 mm long 17

17. Thallus lobes closely appressed to substrate, lower surface ecorticate ....
.... Caloplaca trachyphylla (p. 45)
17. Thallus lobes less tightly appressed to substrate and sometimes ascending away from substrate, lower surface corticate .... Xanthoria elegans (p. 133)

18. Asci polysporous (> 100 spores/ascus), containing tiny, simple, globose to cylindrical hyaline spores... 19
18. Asci not polysporous, containing mostly 8 or fewer spores of various shapes and sizes 22

19. Thallus completely endolithic and indistinct; apothecia gyrose with a carbonized epihymenium .... Polysporina simplex (p. 103)
19. Thallus not endolithic, distinct; apothecia not gyrose, without a carbonized epihymenium 20

20. Cortex C+ red . . . . *Acarospora obpallens* (p. 23)

20. Cortex C– red 21

21. Thalli characterized by areoles that are rounded and mostly consumed by dark brown to black apothecia; thalli epruinose; rare at White Rocks. . . . .
. . . . *Acarospora* sp. nov. (p. 29)

21. Thalli characterized by radial fissures containing small apothecia that are reddish-brown; thalli ranging from densely pruinose and thus nearly white in color to completely epruinoise and thus brown in color, even across a single thallus; extremely common at White Rocks . . . . *Acarospora strigata* (p. 27)

22. Thallus completely endolithic or lacking 23

22. Thallus conspicuous, not endolithic or lacking. 24

23. Margins of apothecia conspicuously cracked and containing usnic acid; thallus ranging from yellow or pale green to white in color; hypothecium with birefringent crystals . . . . *Lecanora* sp. nov. (p. 71)

23. Margins of apothecia not conspicuously cracked and lacking usnic acid; thallus never a shade of yellow or green in color; hypothecium without birefringent crystals . . . . *Lecanora flowersiana* (p. 65)

24. Thallus foliose or squamulose, with a clearly differentianted lower cortex and upper cortex, not tightly appressed to substrate and thus relatively easy to remove from substrate. 25

24. Thallus crustose, lacking a lower cortex, tightly appressed to substrate and thus not easily removable intact 26

25. Thallus squamulose, squamules ≤ 5 mm wide, slightly imbricate, often distinctly pruinose along the margins. . . . . *Psora tuckermanii* (p. 105)

25. Thallus foliose, lobes usually > 5 mm wide, not imbricate, never pruinose along the margins. 26

26. Apothecia brown; thallus not umbilicate. . . . .
. . . . *Xanthoparmelia coloradoensis* (p. 129)

26. Apothecia orange; thallus umbilicate . . . . *Rhizoplaca chrysoleuca* (p. 109)

27. Spores muriform. 28

27. Spores simple or transversely septate but never muriform 29

28. Thallus continuous, chalky white to light gray in color, apothecia round, sunken, and crater-like, sometimes appearing superficiallly similar to perithecia, C+ red cortex, K– epihymenium, asci 8-spored . . . . . Diploschistes scruposus (p. 55)
28. Thallus areolate, brownish-purple in color, apothecia angular, more or less planar with areoles, black, C– cortex, K+ purple epihymenium, asci 1-spored . . . . Rhizocarpon disporum (p. 107)

29. Spores transversely septate, brown 30
29. Spores simple, hyaline 32

30. Spores 4-celled . . . . Diplotomma venustum (p. 57)
30. Spores 2-celled 31

31. Thallus and areoles thick, sordid white to light gray, sometimes stalked in appearance . . . . Rinodina straussii (p. 113)
31. Thallus and areoles thin, dark gray, not stalked in appearance . . . .
. . . . Rinodina venostana (p. 115)

32. Apothecia with lecanorine margins 33
32. Apothecia with lecideine margins 37

33. Thallus crustose, areoles without lobes . . . . Aspicilia cinerea (p. 31)
33. Thallus placidioid, areoles clearly lobed at the margins 34

34. Thallus lobes short and stubby, ~1 mm long, with atranorin (TLC required for all placidioid Lecanoras) . . . . Lecanora argopholis (p. 63)
34. Thallus lobes elongate, always > 2 mm long, lacking atranorin 35

35. Thallus flattened, lobes closely appressed to substrate, usually with leucotylin
. . . . Lecanora muralis (p. 69)
35. Thallus more three-dimensional than above, lobes convex or crinkled in appearance, lacking leutcotylin 36

36. Thallus lobes crinkled in appearance, with usnic acid, K– . . . .
. . . . Lecanora garovaglii (p. 67)
36. Thallus lobes convex but not crinkled, upper cortex and medulla with norstictic acid (K+ red) . . . . Lobothallia alphoplaca (p. 91)

46. Rhizinomorphs absent from lower cortex .. Dermatocarpon americanum (p. 51)
46. Rhizinomorphs present on lower cortex ...... Dermatocarpon moulinsii (p. 53)

47. Thallus squamulose; margins of squamules generally densely white pruinose .... Placidium squamulosum (p. 99)
47. Thallus crustose. 48

48. Areoles subdivided; perithecia not restricted to margins. ....
.... Verrucaria glaucovirens (p. 123)
48. Areoles contiguous; perithecia +/− restricted to margins. ....
.... Verrucaria beltraminiana (p. 119)

49. Thallus leprose, consisting entirely of small, ecorticate granules. ....
.... Lepraria finkii (p. 85)
49. Thallus foliose or crustose 50

50. Thallus K+ violet (anthraquinones present) 51
50. Thallus K− (anthraquinones absent) 52

51. Thallus yellowish-orange; apothecia, when present, orange ....
.... Caloplaca decipiens (p. 35)
51. Thallus gray; apothecia, when present, black. .... Caloplaca pratensis (p. 37)

52. Thallus crustose. .... Verrucaria furfuracea (p. 121)
52. Thallus foliose 53

53. Thallus isidiate, yellowish-green .... Xanthoparmelia lavicola (p. 131)
53. Thallus sorediate, olive brown or brownish-gray 54

54. Thallus gray in color; lobes small (~1 mm wide) and closely appressed to substrate .... Phaeophyscia nigricans (p. 97)
54. Thallus olive brown in color; lobes relatively large (mostly > 1 mm wide) and often lifting above substrate 55

55. Medulla C+ red (gyrophoric acid present); pseudocyphellae +/− conspicuous .... Montanelia tominii (p. 95)
55. Medualla C− red (gyrophoric acid absent); pseudocyphellae often inconspicuous .... Montanelia disjuncta (p. 93)

Abridged Glossary

of Lichenological Terms Used in This Field Guide

apothecium (pl. apothecia). In sexually reproductive species, the "fruiting bodies" of lichens in which asci containing spores derived from meiosis occur.

areole (or "areolate"). An irregular piece of thallus, usually small in size; term is applied to crustose lichens, and areoles are physically separated in space from one another as opposed to similar growth forms not referenced in this guide (e.g., rimose thalli); *Acarospora obpallens* has wonderful areoles.

ascus (pl. asci). The sac-like structure characteristic of all Ascomycete fungi in which sexual reproduction (meoisis followed generally by one or more rounds of mitosis) occurs.

asexual. In reference to an asexually reproducing lichen, through soredia or isidia or similar means (versus other species that reproduce sexually).

biotic soil crust. In reference to any number of species that contribute myriad and significant ecological functions to their environments, particularly arid landscapes; at White Rocks, BSCs help to stabilize loose and fragile soils and sandstone.

bullate. Rounded or spherical areolae that appear inflated; *Lecidella carpathica* is an excellent example of a bullate thallus.

calcareous. In reference to substrates (rock or soil derived from rock) that are rich in carbonates and generally have a high pH, such as limestone.

complex. What you would have if you were composed of ten or twenty different species that all look the same yet somehow are different; typically used in the context of a "species complex," meaning definitely more than one taxonomic entity in a group that has only a single name. Think: *Lecidea tessellata*.

congener. A different species in the same genus (e.g., *Arthonia kermesina* and *Arthonia susa* are congeners).

continuous. In reference to a thallus of a crustose lichen that lacks cracks or some other means of spatial separation between portions of the thallus, which would make it an areolate (instead of continuous) thallus; *Lecidella patavina* has continuous (to areolate) thalli.

cortex (pl. cortices). The uppermost or lowermost (i.e., upper cortex and lower cortex) layers of a lichen, as seen in cross section, composed of highly organized fungal hyphae; crustose, foliose, and squamulose lichens have an upper cortex, whereas only foliose and some squamulose lichens have a lower cortex (fruticose lichens have only one cortex, with upper and lower surfaces undistinguishable).

corticolous. Growing on bark of a living plant (versus lignicolous).

crustose. Lichens that are very closely appressed to the substrate and, by definition, lack a lower cortex; thalli of crustose lichens range from continuous to areolate to squamulose and everything in between; crustose lichens are sometimes called "microlichens." Some people think they are more difficult to identify than macrolichens, but not you.

cyanolichen. In reference to lichens that have a cyanobacterium versus (one or more) green algae for a photobiont.

disc. An informal term for the exterior surface of the apothecium that occurs within the margins (if present) of an apothecium and is often characteristically colored something worth mentioning.

effigurate. Having a thallus with conspicuous lobes, as in *Caloplaca trachyphylla* and several other lichens at White Rocks.

endolithic. Thalli that are contained beneath the surface of the rock substrate and thus are not visible to the naked eye.

epihymenium. The uppermost layer of the hymenium, the color of which can be diagnostic in some species.

epruinose. Lacking pruina.

exciple. The rim around the hymenium.

fertile. In reference to a sexually reproducing species of lichen (versus other species that reproduce asexually).

foliose. A macrolichen that is more or less leaf-like in shape (flat, sprawling) in which the upper and lower cortex are clearly differentiated from one another.

fruiting body. Informal for "apothecium" or "perithecium" (or in reference to sexually reproducing lichens versus asexual species).

fruticose. A macrolichen that is more or less three-dimensional in shape in which the upper and lower cortex are not differentiated from one another.

gyrose (referring to apothecia). Sinuous and turning, as in the discs of *Polysporina simplex*.

hyaline. A common descriptor of spores (e.g., "hyaline" versus "brown") and other vegetative or reproductive features of lichens; meaning transparent or clear. Perhaps Nabokov would have preferred "diaphanous."

hymenium. The fertile portion of an apothecium (or fruiting body) that contains many asci, which in turn contain ascospores.

hypothecium. A distinctive layer of tissue below the hymenium, which can be diagnostically colored in some species; the hypothecium is not immediately adjacent to the spore-bearing hymenium but is instead removed from it by a single layer known as the subhymenium (not used in this guide), which consists of hyphae that give rise to asci.

imbricate. Overlapping, as in shingles on a roof. Just be glad I didn't give you incubous and succubous, too. Let's leave that to the bryologists.

inspersed. In reference to an object that obscures some part of the lichen. For example, when oil droplets obscure the hymenium, it is said to be inspersed.

involucrellum (pl. involucrella). A thickened and morphologically distinctive layer of hyphae surrounding the ostiole of a perithecium. Sounds pretty normal, right?

isidium (pl. isidia). Asexual reproductive structures of some lichens that disperse both the mycobiont and the photobiont and are columnar in shape to some degree (see "soredia" in contrast).

labriform. Lip-like, as in soralia that occur along thallus margins or the undersides of lobes. Can also be used to describe lichen fruiting bodies, as in the genus *Arthonia* (wow!).

laminal. In reference to structures such as apothecia, soredia, or isidia that occur on the main surface of thalli (versus structures that occur only on the margins of thalli).

lecanorine. Margins of the apothecia that are thalline and contain algae (versus lecideine); an artificial division within lichens but a useful feature for dichotomous keys.

lecideine. Margins of an apothecium that lack algae (versus lecanorine); an artificial division within lichens but a useful feature for dichotomous keys.

lichen. A symbiotic organism between a minimum of one fungus and one alga but almost always much more of a complex microcosm of microbial life, with generally a primary mycobiont and a primary photobiont but many additional secondary symbionts. In other words, life at its finest.

lichenicolous. Species of fungi, lichenized or non-lichenized, that specialize in growing on other lichens; many but not all lichenicolous species are parasitic.

lignicolous. Growing on wood of a dead tree or stump, including decaying wood (versus corticolous).

lobe (or "lobule"). An outgrowth of the main thallus of foliose species; if lichens were trees, we would call these "branches"; if lichens were leaves, we would call these "lobes."

macrolichen. Generally large lichens that sprawl away from the substrate or closely appressed (as in crustose lichens) but, by definition, always have a lower cortex; sometimes the lower cortex is not differentiable from the upper cortex, as is the case for most fruticose species.

margin. The tissue surrounding an apothecium, which can contain only fungal hyphae (in which case the margin is "lecideine") or can contain fungal hyphae plus algae (in which case the margin is "lecanorine"); margin can also refer simply to other aspects of a lichen, such as margins of lobes. Or on the margin of society.

medulla. A layer of the lichen thallus composed only of loosely packed fungal hyphae, which occurs below the photobiont layer and is typically (but not always) white in color.

monophyletic. In reference to a group of species that forms a "natural" lineage, that is, one in which all species are each others' closest relatives and no members of this group have been left out; in other words, an ancestor and all of its descendants. A concept that keeps many a systemist employed.

muriform. Spores (sexual products of meiosis) that are highly divided/partitioned morphologically, not just along the long axis (which would result in transversely septate spores) but also along the short axis; consider a series of chambers that can be stacked either end to end ("transversely septate") or end to end as well as side to side ("muriform"). In other words, having horizontal as well as vertical septa.

mycobiont. The (primary) fungal partner of the lichen symbiosis, upon which lichen taxonomy is based, at least for now (many other fungal symbionts are often present, but these are secondary, not primary . . . supposedly).

paraphyses. Non-reproductive fungal hyphae that fill portions of the hymenium not occupied by asci. These can be branched, pigmented, or otherwise ornamented in some manner, which often has taxonomic importance. Think of these as sterile packing tissue.

paraplectenchymatous. A superficial way of thinking about this term is when long strands of fungal tissue are arranged in such a way that it looks like staring through the ends of a giant pile of straws.

photobiont. The green algal or cyanobacterial partner of the lichen symbiosis; we used to think that most species of lichens have only one photobiont, but now we are not so sure. In fact, it appears as though many to most lichens have multiple different photobionts.

prosoplectenchymatous. A superficial way of thinking about this term is when long strands of fungal tissue are arranged in such a way that it looks like peering into a box of straws, not standing on end but with ends arranged to the left and right of the box; the opposite of paraplectenchymatous. If that doesn't work, then think about a box of Q-tips, instead.

prothallus (pl. prothalli). A fungus-only layer that underlies and gives rise to the main thallus; prothalli are usually either white or dark in color and are not visible on all lichens.

pruina. Dead fungal cells usually whitish in color that, where present, give a powdery appearance to the upper surfaces and apothecia discs of lichens. Seeing pruina makes me want to eat funnel cakes, until I think a bit harder and remember how sick those things used to make me feel.

pseudocyphella (pl. pseudocyphellae). Cracks in the upper or lower cortex that expose the medulla. I know, it sounds sort of embarrassing.

rhizine. Differentiated hyphae that arise from a lower cortex and serve to attach some lichens to their substrates; usually associated with foliose lichens.

rhizinomorphs. Rhizine-like structures that similarly arise from lower cortices but do not serve an attachment function; typical of many species in *Umbilicaria* and *Dermatocarpon*. Are these evolutionary holdovers?

rugose. In reference to surfaces of thalli that are wrinkled in appearance; *Dermatocarpon* is often rugose.

saxicolous. Growing on rock.

septum (pl. septa). Partitions or cross walls within spores or hyphae of fungi (where present, said to be "septate"); species of *Buellia* have septate spores, whereas species of *Lecidea* do not.

sexual. In reference to lichens that reproduce through sexual modes and undergo meiosis to produce spores (versus other species that reproduce asexually and undergo mitosis to produce propagules).

silicaceous. In reference to nutrient-poor rock that is generally rich in silica.

simple. When spores are not septate, they are said to be simple; also refers to other aspects of lichens such as "simple rhizines" (when rhizines are unbranched) or "simple apothecia" (when apothecia are not clustered).

sinuous. Wavy anything, just as in regular English-language usage.

soralium (pl. soralia). Specialized structures in which soredia form.

soredium (pl. soredia). Asexual reproductive propagules of some lichens that disperse both the mycobiont and the photobiont and are granular or spherical in shape (see "isidia" in contrast).

specific epithet. One of three required parts of a full name of a species in botany, mycology, and lichenology. The species name consists of a genus, a specific epithet, and a taxonomic authority. The specific epithet is an adjective that describes the genus, which is a noun. With respect to the name *Candelariella clarkiae* E. Tripp and Lendemer, *Candelariella* is the genus, *clarkiae* is the specific epithet, and E. Tripp and Lendemer is the taxonomic authority.

spore (or "ascospore"). One or more products of meiosis (often followed by mitosis) that serve as sexual propagules for lichen reproduction (this process of sexual reproduction refers only to the mycobiont, not the photobiont).

squamulose. Composed of shingle-like areoles in which some portion of the areole often ascends away from the substrate; squamules can be widely separated in space, as in those of *Placidium squamulosum*, or tightly packed together, as in those of *Acarospora stapfiana*.

terricolous. Growing on the ground, as in species on soil.

thallus (pl. thalli). The body of a lichen upon which sexual reproductive structures (e.g., apothecia, perithecia) or asexual reproductive structures (e.g., soredia, isidia) can be found.

tholus. Tips of asci that are generally thicker than remaining portions and whose I+ reactions can be important taxonomically.

transverse (or "transversely septate"). Spores (sexual products of meiosis) that are divided/partitioned morphologically only along the long axis, versus spores that are partitioned along the long as well as short axis; consider a series of chambers that can be stacked either end to end ("transversely septate") or end to end as well as side to side ("muriform").

umbilicate. In reference to species that are attached to the substrate through a single central holdfast, as in the genus *Umbilicaria*.

verruculose. Warty in appearance; *Lecanora argopholis* is an excellent example of a verruculose thallus, as are thalli of the genus *Verrucaria* (surprise!).

voucher. The museum specimen that represents a field collection; without a voucher, science is non-repeatable; vouchers are the amazingly informative and powerful specimens that comprise the world's natural history museums and that represent biodiversity on planet Earth as we know it.

Literature Cited

Amtoft, A., F. Lutzoni, and J. Miadlikowska. 2008. "*Dermatocarpon* (Verrucariaceae) in the Ozark Highlands, North America." *Bryologist* 111 (1): 1–40. http://dx.doi.org/10.1639/0007-2745(2008)111[1:DVITOH]2.0.CO;2.

Anderson, R. A. 1962. "The Lichen Flora of the Dakota Sandstone in North-Central Colorado." *Bryologist* 65 (3): 242–65. http://dx.doi.org/10.1639/0007-2745(1962)65[242:TLFOTD]2.0.CO;2.

Arnold, A. E., J. Miadlikowska, K. L. Higgins, S. D. Sarvate, P. Gugger, A. Way, V. Hofstetter, F. Kauff, and F. Lutzoni. 2009. "A Phylogenetic Estimation of Trophic Transition Networks for Ascomycetous Fungi: Are Lichens Cradles of Symbiotrophic Fungal Diversification?" *Systematic Biology* 58 (3): 283–97. http://dx.doi.org/10.1093/sysbio/syp001.

Arup, U., U. Sochting, and P. Froden. 2013. "A New Taxonomy of the Family Teloschistaceae." *Nordic Journal of Botany* 31 (1): 16–83. http://dx.doi.org/10.1111/j.1756-1051.2013.00062.x.

Bergamini, A., S. Stofer, J. Bolliger, and C. Scheidegger. 2007. "Evaluating Macrolichens and Environmental Variables as Predictors of the Diversity of Epiphytic Microlichens." *Lichenologist* 39 (5): 475–89. http://dx.doi.org/10.1017/S0024282907007074.

Blanco, O., A. Crespo, P. K. Divakar, T. L. Esslinger, D. L. Hawksworth, and H. T. Lumbsch. 2004. "Melanelixia and Melanohalea, Two New Genera Segregated from Melanelia (Parmeliaceae) Based on Molecular and Morphological Data." *Mycological Research* 108 (8): 873–84. http://dx.doi.org/10.1017/S0953756204000723.

Bobbink, R., K. Hicks, J. Galloway, T. Spranger, R. Alkemade, M. Ashmore, M. Bustamante, S. Cinderby, E. Davidson, F. Dentener, et al. 2010. "Global Assessment of Nitrogen Deposition Effects on Terrestrial Plant Diversity: A Synthesis." *Ecological Applications* 20 (1): 30–59. http://dx.doi.org/10.1890/08-1140.1.

Braun, E. L. 1950. *Deciduous Forest of Eastern North America*. Philadelphia: Blakiston.

Breuss, O. 2007a. "Endocarpon." In *Lichen Flora of the Greater Sonoran Desert Region*, vol. 1. ed. T. H. Nash, III, B. D. Ryan, C. Gries, and F. Bungartz, 181–87. Tempe: Lichens Unlimited, Arizona State University.

Breuss, O. 2007b. "Placidium." In *Lichen Flora of the Greater Sonoran Desert Region*, vol. 1. ed. T. H. Nash, III, B. D. Ryan, C. Gries, and F. Bungartz, 384–93. Tempe: Lichens Unlimited, Arizona State University.

Breuss, O. 2007c. "Verrucaria." In *Lichen Flora of the Greater Sonoran Desert Region*, vol. 3. ed. T. H. Nash, III, C. Gries, and F. Bungartz, 335–77. Tempe: Lichens Unlimited, Arizona State University.

Brodo, I. M., S. D. Sharnoff, and S. Sharnoff. 2001. *Lichens of North America*. New Haven, CT: Yale University Press.

Bungartz, F., A. Nordin, and U. Grube. 2007. [2008]. "Buellia." In *Lichen Flora of the Greater Sonoran Desert Region*, vol. 3. ed. T. H. Nash, III, B. D. Ryan, C. Gries, and F. Bungartz, 113–79. Tempe: Lichens Unlimited, Arizona State University.

Byars, L. F. 1936. "An Ecological Study of the Ants of Boulder County, Colorado." MA thesis, University of Colorado, Boulder.

Cameron, R. 2009. "Are Non-Native Gastropods a Threat to Endangered Lichens?" *Canadian Field-Naturalist* 123 (2): 169–71.

Clark, D. A., C. Crawford, and W. F. Jennings. 2001. "Draft Baseline Plant Survey of White Rocks and Surrounding Area in Eastern Boulder County." Unpublished report prepared for the City of Boulder Open Space and Mountain Parks Department.

Conquista, A., A. H. Holmgren, N. H. Holmgren, J. L. Reveal, and P. K. Holmgren. 2013. *Vascular Plants of the Intermountain West U.S.A.* vols. 1–6. Intermountain Flora Full Set. New York: New York Botanical Garden.

Cornelissen, J. H., S. I. Lang, N. A. Soudzilovskaia, and H. J. During. 2007. "Comparative Cryptogam Ecology: A Review of Bryophyte and Lichen Traits That Drive Biogeochemistry." *Annals of Botany* 99 (5): 987–1001. http://dx.doi.org/10.1093/aob/mcm030.

De Wit, T. 1983. "Lichens as Indicators of Air Quality." *Environmental Monitoring and Assessment* 3 (3–4): 273–82. http://dx.doi.org/10.1007/BF00396221.

Divakar, P. K., R. Del-Prado, H. T. Lumbsch, M. Wedin, T. L. Esslinger, S. D. Leavitt, and A. Crespo. 2012. "Diversification of the Newly Recognized Lichen-Forming Fungal Lineage *Montanelia* (Parmeliaceae, Ascomycota) and Its Relation to Key Geological and Climatic Events." *American Journal of Botany* 99 (12): 2014–26. http://dx.doi.org/10.3732/ajb.1200258.

Esslinger, T. L. 1977. "A Chemosystematic Revision of the Brown *Parmeliae*." *Journal of the Hattori Botanical Laboratory* 42:1–211.

Esslinger, T. L. 1986. "Studies in the Lichen Family Physiaceae. VII. The New Genus *Physciella*." *Mycologia* 78 (1): 92–97. http://dx.doi.org/10.2307/3793382.

Esslinger, T. L. 2007. "Phaeophyscia." In *Lichen Flora of the Greater Sonoran Desert Region*, vol. 2. ed. T. H. Nash, III, B. D. Ryan, C. Gries, and F. Bungartz, 403–14. Tempe: Lichens Unlimited, Arizona State University.

Esslinger, T. L. 2015. "A Cumulative Checklist for the Lichen-Forming, Lichenicolous and Allied Fungi of the Continental United States and Canada." North Dakota State University, Fargo. First posted December 1, 1997, most recent version (#20) April 19, 2015. http://www.ndsu.edu/pubweb/~esslinge/chcklst/chcklst7.htm.

Feuerer, T., and E. Timdal. 2007. "Rhizocarpon." In *Lichen Flora of the Greater Sonoran Desert Region*, vol. 2. ed. T. H. Nash, III, B. D. Ryan, C. Gries, and F. Bungartz, 456–66. Tempe: Lichens Unlimited, Arizona State University.

Flora of North America Editorial Committee, ed. 1993. *Flora of North America North of Mexico.* 19+ vols. New York: Oxford University Press.

Freebury, C. E. 2014. "Lichens and Lichenicolous Fungi of Grassland National Park (Saskatchewan, Canada)." *Opuscula Philolichenum* 13:102–21.

Gaya, E. 2009. "Taxonomic Revision of the *Caloplaca saxicola* Group (Teloschistaceae, Lichen-Forming Ascomycota)." *Bibliotheca Lichenologica* 101:1–191.

Hawksworth, David L. 1991. "The Fungal Dimension of Biodiversity: Magnitude, Significance, and Conservation." *Mycological Research* 95 (6): 641–55. http://dx.doi.org/10.1016/S0953-7562(09)80810-1.

Jarman, S. J., and G. Kantvilas. 1995. *A Floristic Study of Rainforest Bryophytes and Lichens in Tasmania's Myrtle-Beech Alliance.* Canberra, Australia: Forestry Tasmania and Department of the Environment, Sport and Territories.

Jepson Flora Project, ed. 2016. *Jepson eFlora.* Accessed July 10, 2016. http://ucjeps.berkeley.edu/eflora.

Kane, Nolan, Erin Tripp, James Lendemer, and Christy McCain. In progress. "Biodiversity Gradients in Obligate Symbiotic Organisms: A Case Study in Lichens in a Global Hotspot." Dimensions of Biodiversity NSF grant.

Kantvilas, G. 1990. "Succession in Rainforest Lichens." *Tasforests* 2: 167–71.

Knoph, J.-G. 1990. "Untersuchungen an gesteinsbewohnenden xanthonhaltigen Sippen der Flechtengattung Lecidella (Lecanoraceae, Lecanorales) unter besonderer Berücksichtigung von außereuropäischen Proben exklusive Amerika." *Bibliotheca Lichenologica* 36:1–183.

Knoph, J.-G., and C. Leuckert. 2007. "Lecidella." In *Lichen Flora of the Greater Sonoran Desert Region*, vol. 2. ed. T. H. Nash, III, B. D. Ryan, C. Gries, and F. Bungartz, 309–20. Tempe: Lichens Unlimited, Arizona State University.

Knudsen, K. 2007. "*Acarospora*." In *Lichen Flora of the Greater Sonoran Desert Region*, vol. 3. ed. T. H. Nash, III, B. D. Ryan, C. Gries, and F. Bungartz, 1–38. Tempe: Lichens Unlimited, Arizona State University.

Knudsen, K. 2007. "Polysporina." In *Lichen Flora of the Greater Sonoran Desert Region*, vol. 3. ed. T. H. Nash, III, B. D. Ryan, C. Gries, and F. Bungartz, 276–78. Tempe: Lichens Unlimited, Arizona State University.

Ladd, D. 1998. "Looking at Lichens." *Missouri Conservationist* 59: 8–13.

Lendemer, J. C. 2013. "A Monograph of the Crustose Members of the Genus *Lepraria* Ach. s. str. (Stereocaulaceae, Lichenized Ascomycetes in North America North of Mexico)." *Opuscula Philolichenum* 11:27–141.

Lendemer, J. C., R. C. Harris, and E. A. Tripp. 2013. *Lichens and Lichenicolous Fungi of Great Smoky Mountains National Park. Memoirs of the New York Botanical Garden*. New York: New York Botanical Garden Press.

Lendemer, J. C., E. A. Tripp, and J. Sheard. 2014. "Review of Rinodina Ach. in the Great Smoky Mountains Highlights the Significance of This 'Island of Biodiversity' in North America." *Bryologist* 117:259–81. http://dx.doi.org /10.1639/0007-2745-117.3.259.

Lindblom, L. 2006. "*Xanthomendoza galericulata*, a New Sorediate Lichen Species, with Notes on Similar Species in North America." *Bryologist* 109 (1): 1–8. http://dx.doi.org/10.1639/0007-2745(2006)109[0001:XGANSL]2.0.CO;2.

Lindblom, L. 2007. "Xanthoria." In *Lichen Flora of the Greater Sonoran Desert Region*, vol. 2. ed. T. H. Nash, III, B. D. Ryan, C. Gries, and F. Bungartz, 605–11. Tempe: Lichens Unlimited, Arizona State University.

Lutzoni, F., and J. Miadlikowska. 2009. "Lichens: Quick Guide." *Current Biology* 19 (13): R502–3. http://dx.doi.org/10.1016/j.cub.2009.04.034.

McCune, B., J. Dey, J. Peck, K. Heiman, and S. Will-Wolf. 1997. "Regional Gradients in Lichen Communities of the Southeast United States." *Bryologist* 100 (2): 145–58. http://dx.doi.org/10.1639/0007-2745(1997)100 [145:RGILCO]2.0.CO;2.

McCune, B., P. Rogers, A. Ruchty, and B. Ryan. 1998. "Lichen Communities for Forest Health Monitoring in Colorado, USA." Report to the USDA Forest Service, Forest Inventory and Analysis Program, Ogden, UT.

Miadlikowska, J., F. Kauff, F. Högnabba, J. C. Oliver, K. Molnár, E. Fraker, E. Gaya, Josef Hafellner, Valérie Hofstetter, Cécile Gueidan, et al. 2014. "A Multigene Phylogenetic Synthesis for the Class Lecanoromycetes (Ascomycota): 1307 Fungi Representing 1139 Infrageneric Taxa, 317 Genera, and 66 Families." *Molecular Phylogenetics and Evolution* 79:132–68. http://dx.doi.org/10.1016/j.ympev.2014.04.003.

Muggia, L., L. Vancurova, P. Skaloud, O. Peksa, M. Wedin, and M. Grube. 2013. "The Symbiotic Playground of Lichen Thalli—a Highly Flexible Photobiont Association in Rock-Inhabiting Lichens." *FEMS Microbiology Ecology* 85 (2): 313–23. http://dx.doi.org/10.1111/1574-6941.12120.

Nash, T. H., III, ed. 2008. *Lichen Biology*, 2nd ed. Cambridge: Cambridge University Press. http://dx.doi.org/10.1017/CBO9780511790478.

Nash, T. H., III, and J. A. Elix. 2007. "Xanthoparmelia." In *Lichen Flora of the Greater Sonoran Desert Region*, vol. 2. ed. T. H. Nash, III, B. D. Ryan, C. Gries, and F. Bungartz, 566–605. Tempe: Lichens Unlimited, Arizona State University.

Nash, T. H., III, B. D. Ryan, P. Diederich, C. Gries, and F. Bungartz, eds. 2004. *Lichen Flora of the Greater Sonoran Desert Region*, vol. 2. Tempe: Lichens Unlimited, University of Arizona.

Nash, T. H., III, B. D. Ryan, P. Diederich, C. Gries, and F. Bungartz, eds. 2007. *Lichen Flora of the Greater Sonoran Desert Region*, vol. 3. Tempe: Lichens Unlimited, University of Arizona.

Nash, T. H., III, B. D. Ryan, C. Gries, and F. Bungartz, eds. 2002. *Lichen Flora of the Greater Sonoran Desert Region*, vol. 1. Tempe: Lichens Unlimited, University of Arizona.

Newmaster, S. G., I. D. Thompson, R. A. D. Steeves, A. R. Rodgers, A. J. Fazekas, J. R. Maloles, R. T. McMullin, and J. M. Fryxell. 2013.

"Examination of Two New Technologies to Assess the Diet of Woodland Caribou: Video Recorders Attached to Collars and DNA Barcoding." *Canadian Journal of Forest Research* 43 (10): 897–900. http://dx.doi.org/10.1139/cjfr-2013-0108.

Nilsson, S. G., U. Arup, R. Baranowski, and S. Ekman. 1995. "Tree-Dependent Lichens and Beetles as Indicators in Conservation Forests." *Conservation Biology* 9 (5): 1208–15. http://dx.doi.org/10.1046/j.1523-1739.1995.9051199.x-ii.

Nordén, B., H. Paltto, F. Götmark, and K. Wallin. 2007. "Indicators of Biodiversity, What Do They Indicate? Lessons for Conservation of Cryptogams in Oak-Rich Forest." *Biological Conservation* 135 (3): 369–79. http://dx.doi.org/10.1016/j.biocon.2006.10.007.

Nordin, A., S. Savic, and L. Tibell. 2010. "Phylogeny and Taxonomy of *Aspicilia* and Megasporaceae." *Mycologia* 102 (6): 1339–49. http://dx.doi.org/10.3852/09-266.

Otálora, M., P. Jørgensen, and M. Wedin. 2014. "A Revised Generic Classification of the Jelly Lichens, Collemataceae." *Fungal Diversity* 64 (1): 275–93. http://dx.doi.org/10.1007/s13225-013-0266-1.

Pettersson, R. B., J. P. Ball, K.-E. Renhorn, P.-A. Esseen, and K. Sjöberg. 1995. "Invertebrate Communities in Boreal Forest Canopies as Influenced by Forestry and Lichens with Implications for Passerine Birds." *Biological Conservation* 74 (1): 57–63. http://dx.doi.org/10.1016/0006-3207(95)00015-V.

Richardson, David H. S. 1999. "War in the World of Lichens: Parasitism and Symbiosis as Exemplified by Lichens and Lichenicolous Fungi." *Mycological Research* 103 (6): 641–50. http://dx.doi.org/10.1017/S0953756298008259.

Ryan, B. D. 2007. "Rhizoplaca." In *Lichen Flora of the Greater Sonoran Desert Region*, vol. 1. ed. T. H. Nash, III, B. D. Ryan, C. Gries, and F. Bungartz, 442–48. Tempe: Lichens Unlimited, Arizona State University.

Ryan, B. D., H. T. Lumbsch, M. I. Messuti, C. Printzen, L. Sliwa, and T. H. Nash, III. 2007. "Lecanora." In *Lichen Flora of the Greater Sonoran Desert Region*, vol. 2. ed. T. H. Nash, III, B. D. Ryan, C. Gries, and F. Bungartz, 176–286. Tempe: Lichens Unlimited, Arizona State University.

Schultz, M. 2005. "An Overview of *Lichinella* in the Southwestern United States and Northwestern Mexico, and the New Species *Lichinella granulosa*."

Bryologist 108 (4): 567–90. http://dx.doi.org/10.1639/0007-2745(2005)10 8[0567:AOOLIT]2.0.CO;2.

Sharnoff, S. 1994. "Use of Lichens by Wildlife in North America." *Research and Exploration* 10: 370–71.

Sheard, J. W. 2010. *The Lichen Genus Rinodina (Ach.) Gray (Lecanoromycetidae, Physciaceae) in North America, North of Mexico.* Ottawa: NRC Research Press.

Showman, R. E. 1987. "Macrolichen Flora of Crane Hollow, Hocking County." *Ohio Journal of Science* 87: 27–29.

Shushan, S., and R. A. Anderson. 1969. "Catalog of the Lichens of Colorado." *Bryologist* 72 (4): 451–83. http://dx.doi.org/10.1639/0007-2745(1969)72[4 51:COTLOC]2.0.CO;2.

Skorepa, A. C. 1973. "Taxonomic and Ecological Studies on the Lichens of Southern Illinois." PhD dissertation. University of Tennessee, Knoxville.

Śliwa, L. 2007. "A Revision of the *Lecanora dispersa* Complex in North America." *Polish Botanical Journal* 52:1–70.

Søchting, U., I. Kärnefelt, and S. Kondratyuk. 2002. "Revision of *Xanthomendoza* (Teloschistaceae, Lecanorales) Based on Morphology, Anatomy, Secondary Metabolites and Molecular Data." *Mitteilungen aus dem Institut für Allgemeine Botanik in Hamburg* 30–32:225–40.

Spribille, T., V. Tuovinen, P. Resl, D. Vanderpool, H. Wolinski, M. C. Aime, K. Schneider, E. Stabentheiner, M. Toome-Heller, G. Thor, H. Mayrhofer, H. Johannesson, and J. P. McCutcheon. 2016. "Basidiomycete Yeasts in the Cortex of Ascomycete Macrolichens." *Science* (July 21). http://science .sciencemag.org/content/early/2016/07/20/science.aaf8287.

Szczepaniak, K., and M. Biziuk. 2003. "Aspects of the Biomonitoring Studies using Mosses and Lichens as Indicators of Methal Pollution." *Enviromental Resources* 93:221–30.

Thomson, J. W. 2007. "Staurothele." In *Lichen Flora of the Greater Sonoran Desert Region*, vol. 1. ed. T. H. Nash, III, B. D. Ryan, C. Gries, and F. Bungartz, 468–72. Tempe: Lichens Unlimited, Arizona State University.

Timdal, E. 1986. "A Revision of *Psora* (Lecideaceae) in North America." *Bryologist* 89 (4): 253–75. http://dx.doi.org/10.2307/3243197.

Tripp, E. A. 2015. "Lichen Inventory of White Rocks Open Space (Boulder,

Colorado)." *Western North American Naturalist* 75: 301–10. http://dx.doi .org/10.3398/064.075.0307.

Tripp, E. A. 2016. "Is Asexual Reproduction an Evolutionary Dead-End in Lichens?" *Lichenologist* 48 (5): 559–80.

Tripp, E. A., and J. C. Lendemer. 2012. "Not Too Late for American Biodiversity?" *Bioscience* 62 (3): 218–19. http://dx.doi.org/10.1525/bio.2012.62.3.2.

Tripp, E. A., and J. C. Lendemer. 2015. "*Candelariella clarkiae* and *Lecidea hoganii*: Two Lichen Species New to Science from White Rocks Open Space, City of Boulder, Colorado." *Bryologist* 118 (2): 154–63. http://dx.doi.org /10.1639/0007-2745-118.2.154.

Weakley, L. S. 2015. *Flora of the Southern and Mid-Atlantic States.* Accessed July 21, 2016. http://www.herbarium.unc.edu/FloraArchives/WeakleyFlora _2015-05-29.pdf (working draft of May 21, 2015).

Weber, W. A. 1949. "The Flora of Boulder County, Colorado." Unpublished report, University of Colorado Museum, Boulder.

Wei, J. C., X. Y. Wang, J. L. Wu, J. N. Wu, X. L. Chen, and J. L. Hou. 1982. *Lichenes Officinales Sinensis.* Beijing: Science Press.

Westberg, M. 2007a. "Candelariella (Candelarielaceae) in Western United States and Northern Mexico: The 8-Spored Lecanorine Species." *Bryologist* 110 (3): 391–419. http://dx.doi.org/10.1639/0007-2745(2007)110[391: CCIWUS]2.0.CO;2.

Westberg, M. 2007b. "Candelariella (Candelarielaceae) in Western United States and Northern Mexico: The Polysporous Species." *Bryologist* 110 (3): 375–90. http://dx.doi.org/10.1639/0007-2745(2007)110[375:CCIWUS]2.0.CO;2.

Westberg, M. 2007c. "Candelariella (Candelarielaceae) in Western United States and Northern Mexico: The Species with Biatorine Apothecia." *Bryologist* 110 (3): 365–74. http://dx.doi.org/10.1639/0007-2745(2007)110[365:CCI WUS]2.0.CO;2.

Westberg, M., C. A. Morse, and M. Wedin. 2011. "Two New Species of Candelariella and a Key to the Candelariales (Lichenized Asomycetes) in North America." *Bryologist* 113 (2): 325–34. http://dx.doi.org/10.1639 /0007-2745-114.2.325.

Wetmore, C. M. 1996. "The *Caloplaca siderites* Group in North and Central America." *Bryologist* 99 (3): 292–314. http://dx.doi.org/10.2307/3244301.

Wetmore, C. M. 2007. "Caloplaca." In *Lichen Flora of the Greater Sonoran Desert Region*, vol. 3. ed. T. H. Nash, III, B. D. Ryan, C. Gries, and F. Bungartz, 179–220. Tempe: Lichens Unlimited, Arizona State University.

Wetmore, C. M. 2009. "New Species of *Caloplaca* (Teloschistaceae) from North America." *Bryologist* 112 (2): 379–86. http://dx.doi.org/10.1639/0007-2745-112.2.379.

Index

hyaline (coloration), 10, 57, 73, 81, 83, 91, 99, 107, 109, 119, 141, 143–44, 147
hydration, 61
hymenium, 61, 65, 73, 79, 81, 99, 117, 139, 144, 147–49
hyphae, 6, 8–9, 11, 61, 146, 148–50
hypothecium, 43, 71, 77, 81, 111, 142, 144, 148

I (iodine), 3–19, 22–139, 141–52
identification, 6, 10, 12, 15–17, 29, 63, 107, 121, 129
images, 15
imbricate, 25, 142, 148
indistinct, 103, 141
infrequent, 109, 135–38
"Ink Pots." See *Rinodina strausii*
insects, 7
inspersed, 79, 81, 148
Interstate 95, 7
inventory assessment, 4, 17
invertebrate, 79
involucrellum (pl. involucrella), 119, 121, 148
"Iron Islands." See *Caloplaca sideritis*
isidium (pl. isidia), 11, 121, 129, 131, 145–46, 148, 151
island, 41, 101

"Jasper Squamules." See *Psora tuckermanii*
Jepson Flora Project, 6

KC test, 23, 25, 27, 31, 33, 35, 37, 39, 41, 43, 45, 47, 49, 51, 53, 55, 57, 59, 61, 63, 65, 67, 69, 71, 73, 75, 77, 79, 81, 83, 85, 87, 89, 91, 93, 95, 97, 99, 101, 103, 105, 107, 109, 111, 113, 115, 117, 119, 121, 123, 125, 127, 129, 131, 133
"K+ Crust." See *Caloplaca pratensis*

labriform, 101, 148
laminal, 93, 101, 148
leaf-like, 12, 147

Lecanora argopholis ("Butter Biscuits"), 63, 91, 136, 143, 152
Lecanora flowersiana ("Seville's Lecanora"), 10, 65, 142
Lecanora garovaglii ("A Textured Dilemma"), 67, 136, 143
Lecanora muralis ("My Old Friend"), 10, 69, 136, 143
Lecanora sp. nov ("Crack Pots"), 13, 63, 65, 67, 69, 71, 91, 109, 136, 142
lecanoric acid, 23, 55, 109
lecanorine, 10, 31, 47, 91, 109, 111, 143, 148–49
Lecidea hoganii ("Timscape"), 10, 16, 47, 57, 73, 81, 136, 139, 144
Lecidea tessellata ("Hornswoggle Lichen"), 75, 77, 81, 136, 139, 144, 146
lecideine, 10, 79, 83, 143, 148–49
Lecidella carpathica ("Lichen Moguls"), 10, 73, 77, 81, 136, 144, 146
Lecidella patavina ("Snail Snacks"), 49, 73, 79. 81, 136, 144, 146
Lecidella stigmatea ("Silver Fox"), 81, 137, 144
Lecidella viridans ("Eggplant in Garlic Sauce"), 83, 137, 144
lens, 33, 140
Lepraria finkii ("Spring Confetti"), 9, 10, 85, 137, 145
leprose, 11, 145
lichen, 1, 3–17, 20, 135, 138, 140–41, 144, 146–51; abundance, 7, 13–14, 39, 45, 77, 103, 135–38; anchor, 9; biology, 8–9, 14–15; biota, revision, 3, 5–6, 8–9, 15–17, 49; body, 8; community, 14; identification, 3, 7, 10–12, 15–17, 138, 147; importance of, 5–6, 17, 79; layers, 8, 11, 146–50; non-sexual portion, 8; species richness, 5, 7, 12, 63; vegetative portion, 8, 147. *See also individual lichen names*
lichenicolous fungus, 25, 33, 115, 148
"Lichen Lettuce." See *Dermatocarpon americanum*

obligate symbiotic relationship, 5
occasional, 37, 53, 65, 75, 89, 109, 135–38
oil, 79, 81, 148
"On the Rocks." See *Verrucaria furfuracea*
Open Space and Mountain Parks (OSMP), 4
"Orange Atoms." See *Caloplaca subsoluta*
"Orange Spectacle." See *Xanthoria elegans*
ostiole, 119, 148
outcroppings, 3, 63
"Outlaw Ashes." See *Verrucaria beltraminiana*
"Over-Easy Lichen." See *Candelariella rosulans*

P test, 23, 25, 27, 31, 33, 35, 37, 39, 41, 43, 45, 47, 49, 51, 53, 55, 57, 59, 61, 63, 65, 67, 69, 71, 73, 75, 77, 79, 81, 83, 85, 87, 89, 91, 93, 95, 97, 99, 101, 103, 105, 107, 109, 111, 113, 115, 117, 119, 121, 123, 125, 127, 129, 131, 133, 136–39
paraphyses, 107, 149
paraplectenchymatous, 43, 97, 150
parasitic, 8, 25, 33, 55, 57, 138, 141, 148
parent rock material, 6, 49
parietin, 35, 39, 43, 45, 125, 127, 133
parietinic acid, 45, 125, 127, 133
particles, 6
peltate, 99
perforations, 9
perithecia, 10, 51, 61, 99, 105, 119, 140, 143–45, 151
perlatolic acid, 93
perspectives, 8
Phaeophyscia nigricans ("Sandy Oysters"), 97, 137, 145
photobiont, 5–6, 8, 99, 140, 147–51
photography, 15, 16
photosynthesis, 8
phyllidia, 11
phylogenetic, 45, 59
Physciella melanchra ("Rib-Ticklers"), 101, 111, 125, 137, 140
physical, 16

pioneer, 25, 35
placidioid, 63, 67, 69, 91, 143
Placidium squamulosum ("Desert Epoxy"), 99, 137, 145, 151
polarilocular, 37, 140
pollutant sequestration, 14
pollution, 7
Polysporina simplex ("Desperate Dots"), 13, 103, 137, 141, 147
polysporous, 29, 31, 47, 103, 141
prairies, 3
preservation, 3, 4–5
primary decomposer, 6
propagules, specialized, 11
prosoplectenchymatous, 97, 150
prothallus (pl. prothalli), 31, 61, 75, 107, 117, 139, 144, 150
protocetraric acid, 129
pruina, 9, 27, 105, 147, 150
pseudocyphella (pl. pseudocyphellae), 9, 12, 93, 95, 145, 150
Psora tuckermanii ("Jasper Squamules"), 9, 10, 105, 137, 142
psoromic acid, 67, 109, 131
public access, 4
"Purple Prose." See *Rhizocarpon disporum*
pycnidia, 11, 129

quartz, 4

rains, 4
rare, 3–4, 7–8, 10, 16, 29, 31, 41, 47, 49, 65, 101, 125, 135–38, 142; animal, 4, 8; plant, 8; species, 4, 7–8, 10, 41, 97, 101, 135–38, 142
ratio, 6
reagent, 15
Red Listed species, 7
regional inventory, 5
removable, 133, 142
report, 5, 16, 33, 115, 135, 139
reproduction: asexual, 11; sexual, 9–11, 35, 37, 129, 140, 144, 146–47, 149, 151–52